Ronny Pini

# Enhanced Coal Bed Methane recovery finalized to carbon dioxide storage

Ronny Pini

# Enhanced Coal Bed Methane recovery finalized to carbon dioxide storage

Experiments and models

Südwestdeutscher Verlag für Hochschulschriften

**Impressum/Imprint (nur für Deutschland/ only for Germany)**
Bibliografische Information der Deutschen Nationalbibliothek: Die Deutsche Nationalbibliothek verzeichnet diese Publikation in der Deutschen Nationalbibliografie; detaillierte bibliografische Daten sind im Internet über http://dnb.d-nb.de abrufbar.
 Alle in diesem Buch genannten Marken und Produktnamen unterliegen warenzeichen-, marken- oder patentrechtlichem Schutz bzw. sind Warenzeichen oder eingetragene Warenzeichen der jeweiligen Inhaber. Die Wiedergabe von Marken, Produktnamen, Gebrauchsnamen, Handelsnamen, Warenbezeichnungen u.s.w. in diesem Werk berechtigt auch ohne besondere Kennzeichnung nicht zu der Annahme, dass solche Namen im Sinne der Warenzeichen- und Markenschutzgesetzgebung als frei zu betrachten wären und daher von jedermann benutzt werden dürften.

Verlag: Südwestdeutscher Verlag für Hochschulschriften Aktiengesellschaft & Co. KG
Dudweiler Landstr. 99, 66123 Saarbrücken, Deutschland
Telefon +49 681 37 20 271-1, Telefax +49 681 37 20 271-0
Email: info@svh-verlag.de
Zugl.: Zürich, ETH, Diss., 2009

Herstellung in Deutschland:
Schaltungsdienst Lange o.H.G., Berlin
Books on Demand GmbH, Norderstedt
Reha GmbH, Saarbrücken
Amazon Distribution GmbH, Leipzig
**ISBN: 978-3-8381-1477-4**


**Imprint (only for USA, GB)**
Bibliographic information published by the Deutsche Nationalbibliothek: The Deutsche Nationalbibliothek lists this publication in the Deutsche Nationalbibliografie; detailed bibliographic data are available in the Internet at http://dnb.d-nb.de.
 Any brand names and product names mentioned in this book are subject to trademark, brand or patent protection and are trademarks or registered trademarks of their respective holders. The use of brand names, product names, common names, trade names, product descriptions etc. even without a particular marking in this works is in no way to be construed to mean that such names may be regarded as unrestricted in respect of trademark and brand protection legislation and could thus be used by anyone.

Publisher: Südwestdeutscher Verlag für Hochschulschriften Aktiengesellschaft & Co. KG
Dudweiler Landstr. 99, 66123 Saarbrücken, Germany
Phone +49 681 37 20 271-1, Fax +49 681 37 20 271-0
Email: info@svh-verlag.de

Printed in the U.S.A.
Printed in the U.K. by (see last page)
**ISBN: 978-3-8381-1477-4**


Copyright © 2010 by the author and Südwestdeutscher Verlag für Hochschulschriften Aktiengesellschaft & Co. KG and licensors
All rights reserved. Saarbrücken 2010

# Acknowledgements

Thanks to Prof. Marco Mazzotti for supervising my research project and for putting me in the condition to work and express myself at best. I learnt a lot in these years and I owe it mainly to him.

Prof. Giuseppe Storti showed since the beginning a lot of interest in my research. He has been always ready to help me on every topic: from mathematical to experimental issues, from polymers to coal. Thank you.

The help of Luigi Burlini has been essential, in particular for introducing me to the geological aspects related to my research, thus allowing me to look at it from a different perspective.

I am very grateful to Prof. Daniel Tondeur for accepting the task of co-examiner and for taking the time to read my thesis.

There would haven't been a single experimental point on the graphs presented in this thesis without the extremely valuable help of the technicians. I am very grateful to Chris Rohrbach for being always ready to help me with technical problems and for sharing its opinion on innumerable topics. Thanks go also to Marcel Mettler and Robert Hofmann from the geological department for their help in the lab.

Olga Williams, Dorian Marx and David Wochner are students who worked with me for their semester- and master projects. Supervising them has been an enjoyable experience, allowing me to learn a lot from the many discussions we had.

I would like to thank all my colleagues for the wonderful working atmosphere. In particular, Stefan Ottiger, with whom I spent most of my PhD, and Christian Langel and Christian Lindenberg for the countless nice moments lived together; I feel happy and lucky when I think at the barbecues in La's garden, at the weekends in Ascona, at Li's culinary art, at the Saturday's lunches in Pentagramma and at the evenings in the Hafenkneipe. Thank you!

There is no good work, without inspiration...and mine is music. Mischa brought with his finest selection of music the perfect working atmosphere in the office and I really enjoyed it.

Finally, special thanks go to my family and my friends for always being there and whose unconditional support made all that possible.

Zurich, May 2009

Ronny Pini

*Un imprevisto è una pausa fra due idee.*
Fabrizio de André

## Abstract

The recovery of coal bed methane can be enhanced by injecting carbon dioxide ($CO_2$) in the coal seam at supercritical conditions. Through an in situ adsorption/desorption process the displaced methane ($CH_4$) is produced and the adsorbed $CO_2$ is permanently stored. This process is called Enhanced Coal Bed Methane recovery (ECBM) and it is a technique under investigation as a possible approach to the geological storage of $CO_2$ in a carbon dioxide capture and storage (CCS) system. ECBM recovery is not yet a mature technology, in spite of the growing number of pilot and field tests worldwide that have shown its potential and highlighted its difficulties. The problems encountered are largely due to the heterogeneous nature of coal and its complex interaction with gases. These issues, which represent the motivation of this research work, need to be addressed both at laboratory and field test scales.

The aim of this thesis is therefore to develop experimental and modeling tools that are able to provide a comprehensive characterization of coal required first to understand the mechanisms acting during the process of injection and storage and secondly to assess its potential for an ECBM operation. In particular, sorption data of $CO_2$, $CH_4$ and $N_2$ on several coals from different coal mines worldwide have been measured at conditions typically encountered in coal seams. $CO_2$ maximum sorption capacity per unit mass of dry coal has been found to range between 5% and 14% weight and to depend on coal rank in a non-monotonic way. Moreover, for a specific coal, competitive sorption isotherms of the binary and ternary mixtures of these gases have been obtained, showing that $CO_2$ adsorbs always more than $CH_4$, and $CH_4$ more than $N_2$. This property is of key importance for ECBM application.

In order to investigate the coal volumetric behavior upon exposure to an adsorbing gas, two approaches have been followed. In the first, the utilization of a visualization technique allowed to measure the unconstrained expansion of a coal disc, confirming that indeed coal swells when exposed to a gas that is able to adsorb on its surface and penetrate into its structure, whereas exposure to an inert gas leads to negligible volume changes. In the second approach, flow experiments on coal cores confined under an external hydrostatic pressure were performed. Moreover, a model describing the fluid transport trough the coal core has been derived, which includes mass balances accounting for gas flow, gas sorption and swelling, and mechanical constitutive equations for the description of porosity and permeability changes during injection. The combination of the experimental data with the model predictions allowed to highlight and to understand the different fundamental aspects of the process dynamics, and to relate them to parameters such as the effective pressure on the sample, adsorption and swelling.

All the outcomes of the above mentioned experimental studies are the needed information to be included in ECBM simulation studies, aimed at the description and design of ECBM processes. A modeling study has been undertaken, where the coal core model used to describe the pure gas injection experiments has been extended to mixtures, thus allowing to investigate the gas flow process in coal beds and the displacement dynamics during an ECBM operation. Particular attention has been given to the injection of $CO_2/N_2$ mixtures (the so-called flue gas), that would allow to avoid the expensive capture step and to keep coal permeability sufficiently high.

## Sommario

L'iniezione di anidride carbonica ($CO_2$) in condizioni supercritiche in strati profondi di carbone permette di facilitare l'estrazione del gas metano ($CH_4$) naturalmente presente nella vena carbonifera. Tramite un processo di adsorbimento/desorbimento il metano è rilasciato e può essere estratto, mentre la $CO_2$, che si lega al carbone, è stoccata in modo permanente. Questa operazione è chiamata Enhanced Coal Bed Methane recovery (ECBM) ed è una delle tecniche proposte per lo stoccaggio geologico della $CO_2$ nel contesto dei sistemi di cattura e stoccaggio di $CO_2$ (Carbon Capture and Storage, CCS). Nonostante il numero crescente di progetti pilota e prove sul campo, che hanno dimostrato il potenziale di questo tipo di operazioni e evidenziato le sue difficoltà, la tecnica ECBM non è ancora abbastanza matura da permettere un suo utilizzo su scala commerciale. Le maggiori difficoltà riscontrate sono dovute soprattutto alla natura eterogenea del carbone e alla sua complessa interazione con i gas. Questi problemi, che rappresentano la motivazione del presente progetto di ricerca, devono essere affrontati sia tramite studi di laboratorio che nuove prove sul campo.

Lo scopo di questa tesi è dunque quello di sviluppare degli strumenti, in termini sia di tecniche di laboratorio che di modellazione, che permettano di fornire una caratterizzazione esauriente del carbone, allo scopo di capire i meccanismi che agiscono durante l'operazione di iniezione e stoccaggio, e di determinare il suo potenziale per un eventuale utilizzo ECBM. In particolare, diversi campioni di carbone provenienti da bacini di carbone da tutto il mondo sono stati analizzati in termini di adsorbimento di $CO_2$, $CH_4$ e $N_2$ in condizoni tipicamente riscontrate in vene carbonifere profonde. La capacità massima di $CO_2$ misurata varia tra il 5% ed il 14% per massa di carbone asciutto e dipende in maniera non-monotona dall'età del carbone. Inoltre, per un carbone specifico sono state ottenute delle isoterme di adsorbimento con miscele binarie e ternarie di questi gas, mostrando che la $CO_2$ si lega maggiormente al carbone rispetto al $CH_4$, e il $CH_4$ a sua volta più del $N_2$. Questo comportamento è l'elemento chiave delle operazioni ECBM.

Il comportamento volumetrico del carbone una volta esposto a dei gas ad alta pressione è stato studiato tramite due metodi diversi. Nel primo, l'utilizzo di una tecnica di visualizzazione diretta ha permesso di misurare l'espansione libera di un disco di carbone confermando che in effetti il carbone si rigonfia se esposto a dei gas come la $CO_2$, il $CH_4$ e l'$N_2$, che adsorbono sulla sua superficie e sono in grado di penetrare nella sua struttura, mentre l'esposizione ad un gas inerte non comporta nessun cambiamento di volume. Con la seconda tecnica sono state effettuate prove di flusso su carote di carbone sottoposte ad una pressione idrostatica esterna. Un modello matematico è stato sviluppato per descrivere il trasporto di gas attraverso la carota di carbone, che include bilanci materiali del flusso di gas, di adsorbimento e di rigonfiamento, ed equazioni meccaniche per descrivere le variazioni di porosità e permeabilità durante le prove di iniezione. La combinazione dei dati sperimentali con le predizioni del modello ha permesso di evidenziare e di capire gli aspetti fondamentali che controllano la dinamica del processo e di relazionarli a parametri come la pressione effettiva esercitata sul campione, l'adsorbimento e il rigonfiamento.

I risultati delle prove sperimentali appena descritte costituiscono un importante pacchetto di informazioni che può essere inserito in simulatori di giacimento, utilizzati per descrivere e pianificare operazioni ECBM. È stato condotto un lavoro di modellizzazione, in cui il modello matematico usato per descrivere le prove di flusso in laboratorio è stato esteso a miscele, permettendo appunto di studiare il trasporto e lo spiazzamento di gas nella vena carbonifera durante un operazione ECBM. Particolare attenzione è stata data all'iniezione di miscele di $CO_2$ e $N_2$ (i cosidetti fumi), che permetterebbe da una lato di evitare il costoso passaggio antecedente l'iniezione, in cui la $CO_2$ è separata dai fumi, e dall'altro di mantenere la permeabilità del carbone sufficientemente alta.

# Table of Contents

**1  Geological storage of $CO_2$ and ECBM recovery** — 1

    1.1  Carbon Capture and Storage — 2

    1.2  Geological storage of $CO_2$ — 3

    1.3  Storage in unmineable coal seams — 5

    1.4  ECBM field tests — 9

    1.5  Research needs — 10

    1.6  Structure of the thesis — 12

**2  Sorption on coal: pure gases** — 15

    2.1  Introduction — 15

    2.2  Experimental — 18

        2.2.1  Coal characterization — 18

        2.2.2  Experimental methods — 19

        2.2.3  Absolute sorption isotherms — 22

|  |  | 2.2.4 | Swelling | 24 |
|---|---|---|---|---|
|  | 2.3 | Sorption isotherm model | | 25 |
|  |  | 2.3.1 | Adsorption and absorption | 25 |
|  |  | 2.3.2 | Proposed model for sorption | 26 |
|  |  | 2.3.3 | Comparison with literature | 29 |
|  | 2.4 | Results | | 32 |
|  |  | 2.4.1 | Comparison among different laboratories | 32 |
|  |  | 2.4.2 | Comparison of different coals | 33 |
|  |  | 2.4.3 | Effect of temperature | 41 |
|  |  | 2.4.4 | Spatial variation of the sorption behavior | 44 |
|  |  | 2.4.5 | Effect of rank | 46 |
|  | 2.5 | Discussion and concluding remarks | | 48 |
|  | 2.6 | Nomenclature | | 52 |
| **3** | **Sorption on coal: gas mixtures** | | | **55** |
|  | 3.1 | Introduction | | 55 |
|  | 3.2 | Experimental section | | 58 |
|  |  | 3.2.1 | Coal characterization | 58 |
|  |  | 3.2.2 | Experimental technique | 58 |
|  |  | 3.2.3 | Absolute sorption | 62 |
|  |  | 3.2.4 | Methodology | 63 |
|  | 3.3 | Results and discussion | | 66 |

|       | 3.3.1 | Binary mixtures . . . . . . . . . . . . . . . . . . . . | 66 |
|-------|-------|---------------------------------------------------------|----|
|       | 3.3.2 | Ternary mixture . . . . . . . . . . . . . . . . . . . . | 76 |
| 3.4   | Concluding remarks . . . . . . . . . . . . . . . . . . . . . . . | | 77 |
| 3.5   | Nomenclature . . . . . . . . . . . . . . . . . . . . . . . . . . | | 79 |

# 4 Swelling of coal — 81

| 4.1 | Introduction . . . . . . . . . . . . . . . . . . . . . . . . . . . | | 81 |
|-----|-------------------------------------------------------------------|---|----|
| 4.2 | Experimental section . . . . . . . . . . . . . . . . . . . . . . | | 84 |
|     | 4.2.1 | Materials and experimental setup . . . . . . . . . . | 84 |
|     | 4.2.2 | Experimental procedure . . . . . . . . . . . . . . . | 85 |
|     | 4.2.3 | Results and discussion . . . . . . . . . . . . . . . | 87 |
| 4.3 | Concluding remarks . . . . . . . . . . . . . . . . . . . . . . . | | 93 |
| 4.4 | Nomenclature . . . . . . . . . . . . . . . . . . . . . . . . . . | | 96 |

# 5 Permeability of coal — 97

| 5.1 | Introduction . . . . . . . . . . . . . . . . . . . . . . . . . . . | | 97 |
|-----|-------------------------------------------------------------------|---|-----|
| 5.2 | Experimental section . . . . . . . . . . . . . . . . . . . . . . | | 100 |
|     | 5.2.1 | Coal sample characterization . . . . . . . . . . . . | 100 |
|     | 5.2.2 | Experimental Setup . . . . . . . . . . . . . . . . . | 104 |
|     | 5.2.3 | Measurement procedure . . . . . . . . . . . . . . . | 105 |
| 5.3 | Modeling . . . . . . . . . . . . . . . . . . . . . . . . . . . . | | 108 |
|     | 5.3.1 | Mass Balances . . . . . . . . . . . . . . . . . . . . | 109 |

      5.3.2   Stress-Strain Relationship . . . . . . . . . . . . . . 111

      5.3.3   Solution Procedure . . . . . . . . . . . . . . . . . . 114

  5.4   Results and Discussion . . . . . . . . . . . . . . . . . . . . 115

      5.4.1   Experiments with an Inert Gas . . . . . . . . . . . 117

      5.4.2   Experiments with an Adsorbing Gas . . . . . . . . 119

      5.4.3   Experiments with closed system . . . . . . . . . . 129

  5.5   Concluding remarks . . . . . . . . . . . . . . . . . . . . . 130

  5.6   Nomenclature . . . . . . . . . . . . . . . . . . . . . . . . . 134

## 6 Coal bed dynamics                                              137

  6.1   Introduction . . . . . . . . . . . . . . . . . . . . . . . . . . 137

  6.2   Modeling . . . . . . . . . . . . . . . . . . . . . . . . . . . 140

      6.2.1   Mass balances . . . . . . . . . . . . . . . . . . . . . 141

      6.2.2   Stress-strain relationship . . . . . . . . . . . . . . 143

      6.2.3   Solution procedure . . . . . . . . . . . . . . . . . 148

  6.3   Results . . . . . . . . . . . . . . . . . . . . . . . . . . . . . 152

      6.3.1   Permeability changes . . . . . . . . . . . . . . . . 152

      6.3.2   ECBM recovery and $CO_2$ storage . . . . . . . . . . 153

  6.4   Discussion and concluding remarks . . . . . . . . . . . . . 164

## 7 Containment in the reservoir                            169

  7.1   Introduction . . . . . . . . . . . . . . . . . . . . . . . . . . 169

|   |   |   |   |
|---|---|---|---|
| | 7.2 | Modeling the flow through the caprock | 175 |
| | | 7.2.1 Diffusion | 175 |
| | | 7.2.2 Convection | 176 |
| | | 7.2.3 Solution Procedure | 178 |
| | | 7.2.4 Results | 181 |
| | | 7.2.5 Discussion | 184 |
| | 7.3 | The gas-in-place in coal seams | 185 |
| | 7.4 | Concluding remarks | 189 |
| | 7.5 | Nomenclature | 190 |
| **8** | **Outlook** | | **193** |
| | 8.1 | Gas sorption on wet samples | 194 |
| | | 8.1.1 Experimental procedure | 196 |
| | | 8.1.2 Results and discussion | 199 |
| | 8.2 | Displacement experiments | 202 |
| **A** | **Proximate analysis of coal** | | **207** |
| | A.1 | Thermogravimetric analysis (TGA) | 207 |
| | A.2 | Experimental procedure | 208 |
| **B** | **List of Figures** | | **211** |
| **C** | **List of Tables** | | **223** |

# Chapter 1

# Geological storage of $CO_2$ and ECBM recovery

Since a few years the debate on climate change is focusing the attention of our society. The uncertainty on both the causes and the effects, and the complexity of this phenomenon keeps the discussion about this topic very lively. Some effects of climate change are easily perceptible: the widespread melting of snow and ice, and the rising global average sea level are the result of the increase in global average air and ocean temperatures. Experts agree that the causes of the warming of the climate system are very likely to be attributed to the increase of greenhouse gases concentration in the atmosphere (IPCC, 2007). Of particular concern are the emissions of carbon dioxide which have increased significantly and exceed by far the pre-industrial values, mainly because of the intense use of fossil fuels, like oil, coal and natural gas (MIT, 2007). Efforts need therefore to concentrate in the drastic reduction of the $CO_2$ emissions, since climate change, if unmitigated, can have serious implications for the economic well-being of human society (IPCC, 2007).

Energy supply is the key to prosperity in the industrialized world and a prerequisite for sustainable development in countries with transition economies. Moreover, the development of energy systems must be accomplished without endangering the quality of life for present and future generations and without exceeding the capacity of supporting ecosystems. The challenge for governments and industry is therefore to find a path that facilitates the achievement of carbon emission reduction goals, while continuing to meet urgent energy needs (MIT, 2007). Several ways may be followed to achieve this goal, namely by reducing the energy consumption, both at the production level through more efficient technologies, and at the consumption level through changes in life habits, by extending the use of zero-$CO_2$ emission technologies such as renewable energies and nuclear energy, and finally by capturing the $CO_2$ produced and storing it deep underground separated from the atmosphere.

## 1.1 Carbon Capture and Storage

Currently 80% of the world's energy demand is covered by fossil sources: coal accounts for 26%, natural gas for 20.5% and oil for 34.4%; only 0.6% of global energy demand is met by geothermal, solar and wind (IEA, 2008). $CO_2$ capture and storage (CCS) has therefore been recognized as the critical enabling technology that would reduce $CO_2$ emissions significantly while also allowing fossil fuels to meet the worlds pressing energy needs (IPCC, 2005; MIT, 2007), during the transition period to the aforementioned zero-emission technologies.

The idea behind CCS is quite simple and it goes through three steps: once captured, for example from a fossil fuel power plant, the $CO_2$ is transported to a location where it can be permanently and safely stored, with the aim of isolating it from the atmosphere. CCS may be first applied where the $CO_2$ is produced in large amounts, i.e. at $CO_2$-intensive industries such as cement production, refineries, oil and gas processing, since one third of $CO_2$ emissions is generated in that manner (IEA, 2008).

Although not yet at the scale required for CCS to be commercially relevant, the first two steps belonging to the CCS chain are mature technologies in the industry. $CO_2$ is routinely separated from gas streams in the natural gas processing as well as in hydrogen production, whereas $CO_2$ transport is a well-known market in the USA, where more than 40 $MtCO_2$ per year are transported over 2500 km of pipelines (IPCC, 2005). The storage of $CO_2$ into natural geological formations would require the use of the same technologies (drilling, compression, etc.) that have been developed by the oil and gas industry in the last decades, with the further requirement that in this case the $CO_2$ has to be permanently trapped in the geological structure and monitored. In addition to that, the main challenge to be addressed is the demonstration of an integrated system of capture, transportation, and storage of $CO_2$ (MIT, 2007).

## 1.2 Geological storage of $CO_2$

Geological disposal of the $CO_2$ is a possible storing method and several geological settings may act as host for the captured $CO_2$, namely depleted oil- and gas fields, deep saline formations and unmineable coal seams. In all cases we are dealing with layers of porous rocks located at around 1 km depth under a so-called cap rock (a layer of impermeable rock such as shale), which act as a seal minimizing the chance of gas leakage. A careful selection of each site has to be carried out, since at those depths the $CO_2$ exists in its supercritical state with roughly half of the density of water and therefore it might tend to migrate upward through the porous overlying strata due to buoyant forces. On the one hand, this migration is prevented and controlled by the low-permeability cap rock and by capillary forces; on the other hand, $CO_2$ can escape if it finds its way through fractures or faults which may be present in the strata surrounding the storage formation (Li et al., 2006; Bachu, 2008). Researches demonstrate that there exists a large potential storage capacity in deep geologic locations around the world, although their distribution is quite uneven (IPCC, 2005). First candidates will be the storage formations near large industrial facilities, which would permit to keep the costs for CCS deployment down. Moreover, first demonstration projects indicate that large-scale $CO_2$ injection (i.e. about 1 $MtCO_2$ per year) can be indeed operated safely, namely the Sleipner project in an off-shore saline formation in Norway (operated by Statoil, Norway), the Weyburn EOR project in Canada (operated by EnCana, Canada), and the In Salah project in a gas field in Algeria (operated by BP, Great Britain). The first of such operations, launched by Statoil, started in 1996: since then approximately 2700 t $CO_2$/day have been injected in a saline formation 800 m below the seabed in the North Sea. The monitoring program, aimed at the verification of the retention of the injected $CO_2$ in the reservoir, successfully confirmed that the cap rock is an effective seal preventing the migration of the $CO_2$ out of the formation. Until 2005, about 7 Mt $CO_2$ have been injected, and within the lifetime of the project, a total of 20 $MtCO_2$ is expected to be stored.

More of such large scale operations are however required, since each reservoir presents unique characteristics that demand site-specific studies. Only the investigation of a range of different geologies will allow improving the knowledge and the confidence in CCS, so that this technology could be implemented at the scale needed to significantly affect the amount of $CO_2$ emitted to the atmosphere in the future.

## 1.3 Storage in unmineable coal seams

Coal seams have been proposed as a possible location for permanent geological storage of carbon dioxide ($CO_2$) (IPCC, 2005). In particular, the coal seams which can be used for storage purposes are those presenting characteristics precluding an economically profitable mining. These so-called unmineable coal seams are either too thin, too deep or too high in sulfur content and mineral matter (White et al., 2005). The estimated storage potential of coal seams is relatively small compared to other geological formations and varies between 3 $GtCO_2$ and 200 $GtCO_2$ (IPCC, 2005). The upper estimate refers to the worldwide distribution of bituminous coal seams, whereas the lower estimate refers only to those coal seams, where simultaneous CBM production could be carried out. These values, together with the distribution of potential coal seams which not always matches the location of large $CO_2$ sources, suggest that the contribution of coal seams to the underground storage of $CO_2$ will be limited compared to other geological formations. This amount is however still significant compared to the current anthropogenic $CO_2$ emissions of almost 30 $GtCO_2$ per year (IPCC, 2007), and need to be taken into account in the effort of finding ways for reducing greenhouse gases emissions.

Coal seams are fractured porous media, characterized by a relatively large internal surface area. Significant amounts of methane ($CH_4$) are generated and retained during the geological process leading to their formation, the so-called coalification process (Levine, 1993; Gentzis, 2000). The way this coal bed methane is stored in the coal reservoir differs from other geological locations in the fact that, besides filling the avail-

# 1. Geological storage of $CO_2$ and ECBM recovery

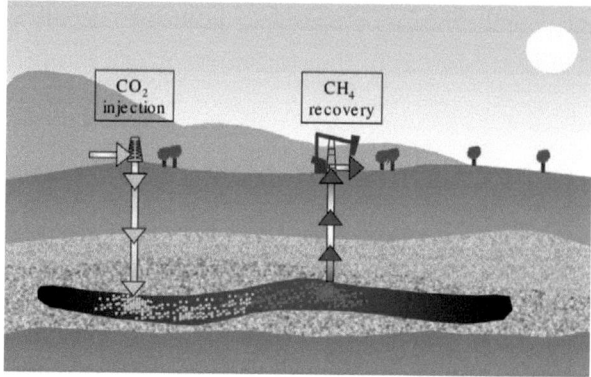

Figure 1.1: Schematic of an ECBM operation, where captured $CO_2$ from a power plant is injected into the coal seam and $CH_4$ is produced. Injection and production wells can in general be more than one.

able fracture and pore volume, the gas adsorbs on the coal surface and absorbs into the coal matrix. In this form it has a much higher density than gas (Sircar, 2001), allowing for a better exploitation of the reservoir rock as a storage medium for $CO_2$.

Such coal bed methane can be recovered from the coal seam and used for energy production. Conventional primary recovery of methane, which is performed by pumping out water and depressurizing the reservoir, allows producing back 20-60% of the methane originally present in the reservoir (White et al., 2005). As in the case of enhanced oil recovery (EOR), such primary production could be in principle enhanced by injecting $CO_2$ in the coal seam. This process is schematized in Figure 1.1 and is called Enhanced Coal Bed Methane (ECBM) recovery (White et al., 2005; Mazzotti et al., 2009). Due to higher adsorptivity of $CO_2$ with respect to $CH_4$, the injected carbon dioxide displaces the adsorbed methane. Ultimately, most of the methane is recovered and the coal seam contains mainly carbon dioxide, which remains there permanently

## 1.3 Storage in unmineable coal seams

separated from the atmosphere.

Once injected underground, $CO_2$ is trapped as a dense gas in the coal cleats, adsorbed on and absorbed in the coal, and solubilized in the formation water. Figure 1.2 shows the conditions to be expected underground in terms of temperature, $CO_2$ density and pressure. Optimal storage conditions are attained at high density, i.e. at a depth of more than 750 m, where pressure is more than 75 bar and temperature is about 40°C or more, and therefore $CO_2$ is supercritical. From an engineering point of view ECBM recovery is thus an adsorption/desorption process at supercritical conditions in a natural underground coal formation, which is accomplished by injecting $CO_2$ in one or more injection wells and by collecting $CH_4$ from one or more production wells.

The process of injecting a gas into a reservoir with simultaneous recovery of a value added product is quite popular in the oil industry, where production of oil is enhanced by injection of $CO_2$ or $N_2$ into the reservoir (EOR). Exactly because of this added value, those techniques that offer a byproduct such as natural gas are expected to be the first commercially practiced storage technologies compared to the other scenarios for long term storage of $CO_2$ where there is no offset of operational costs (White et al., 2005). Moreover, the expertise gained in the past years for enhanced oil production will play an important role in a faster implementation of the ECBM technology at a commercial scale.

In conclusion, ECBM is attractive from two perspectives. On the one hand, if one is interested in the recovered methane as a fuel or a technical gas, ECBM allows also for a net $CO_2$ sequestration, thanks to the above mentioned high $CO_2$ adsorptivity. On the other hand, if the goal is that of storing $CO_2$ that has been captured, the ECBM operation allows also recovering methane, thus making $CO_2$ storage economically interesting in this case.

# 1. Geological storage of CO$_2$ and ECBM recovery

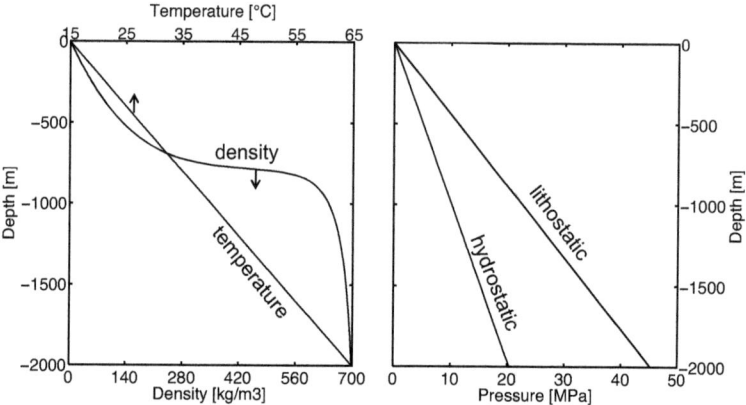

Figure 1.2: CO$_2$ density, pressure and temperature as a function of depth to be expected for underground CO$_2$ storage, assuming a geothermal gradient of 25°C/km from 15°C at the surface, and specific weight of soil and water of 22.62 kN/m$^3$ and 10.18 kN/m$^3$, respectively. CO$_2$ density increases rapidly at 800 m depth, when CO$_2$ reaches its supercritical state. Generally, the hydrostatic pressure is taken as the criterion to determine CO$_2$ injection pressure. The lithostatic pressure is the pressure exerted on the coal bed by the surrounding rock (also called geostatic pressure).

## 1.4 ECBM field tests

Towards demonstration of its feasibility and as a first step in the direction of its commercial deployment, the ECBM technology has been implemented in a number of field tests, which are reported in Table 1.1. The first ECBM project, the Coal-Seq project, was the one at the largest scale and the most meaningful; it took place at the San Juan Basin in New Mexico (USA), where pure $CO_2$ and pure $N_2$ were injected in the Allison and Tiffany Unit, respectively, while $CH_4$ was successfully produced in a multi-well configuration over a period of more than five years (Reeves, 2004). It was shown that gas injection indeed enhanced methane recovery. $CO_2$ injection yielded a reduction in permeability and injectivity, whereas $N_2$ injection led to a much more rapid breakthrough thus reducing product purity. The former effect was attributed to the porosity reduction associated with coal swelling upon $CO_2$ injection, particularly evident near the well, where the $CO_2$ pressure is high. In order to clarify this effect, let us consider the so-called matchsticks model, which describes coal as an ensemble of parallel elongated matrix elements separated by fractures, i.e. cleats, that constitute its transport porosity (Seidle et al., 1992; Gentzis, 2000). On the one hand, the external lithostatic pressure tends to press the matrix elements together, and to reduce porosity (Cui et al., 2007). On the other hand, uptake and release of many gases (absorption) is associated with swelling and shrinking of coal, respectively (Larsen, 2004). Upon gas absorption the coal matrix elements swell, and the space between them is consumed, thus reducing porosity as well. In the case of $N_2$ injection, coal undergoes a net shrinking, since $N_2$ swells less than the displaced $CH_4$, thus enhancing the porosity. Both effects (lithostatic compression and coal shrinking/swelling) lead to a change of the permeability, which needs to be quantified, since permeability controls injection pressure and gas production and affects therefore the overall ECBM operation.

The other field tests in Table 1.1 were on a much smaller scale, exploiting a single well (Gunter et al., 2004; Wong et al., 2006) or a two-well configuration (Van Bergen et al., 2006; Yamaguchi et al., 2006). The goal of these projects was that of testing the ECBM technology in reservoirs with different geological characteristics and to observe $CO_2$ breakthrough within the project life time, usually around 1 year. This information is very useful, in particular when compared to the results obtained from reservoir modeling studies (Van Bergen et al., 2006). In all cases $CH_4$ production was enhanced in response to gas injection and $CO_2$ injection lead to a reduction of injectivity. As in the case of the San Juan Basin, the latter was attributed to the closing of the fracture associated with coal swelling. The low injection rates could be compensated through shut-in periods in the Alberta $CO_2$ ECBM project (Gunter et al., 2004), or through a frac job in the RECOPOL project, at least partially and temporarily (Van Bergen et al., 2006). On the contrary, during the injection of flue gas in the Canadian project a steady increase of well injectivity was observed (Gunter et al., 2004).

## 1.5  Research needs

By testing different gas injection policies, such as $CO_2$, $N_2$ or their mixtures, these field tests have shown the potential of coal seams as storage sites for $CO_2$. However, they also evidenced that many factors affect the success of the ECBM operation, which need to be extensively investigated. The problems encountered are largely due to the heterogeneous nature of coal and the interactions of gases with different structural and chemical features of the coal. These issues, which represent the moti-

## 1.5 Research needs

Table 1.1: ECBM field tests. Well configuration: sw, single-well; 2w, two-well; mw, multi-well.

| Location | Project | Year | Gas | # of wells | Injection | Reference |
|---|---|---|---|---|---|---|
| Fenn-Big Valley (Canada) | Alberta $CO_2$/ECBM | 1999 | $CO_2$ | sw | 0.19 kt | Gunter et al. (2004) |
| | | | 13% $CO_2$ / 87% $N_2$ | sw | 0.11 kt | |
| | | | 53% $CO_2$ / 47% $N_2$ | sw | 0.12 kt | |
| | | | $N_2$ | sw | $\approx 0.1$ kt | |
| South Qinshui Basin (China) | - | 2004 | $CO_2$ | sw | 0.19 kt | Wong et al. (2006) |
| Ishikari Coal Field (Japan) | JCOP | 2004 | $CO_2$ | 2w | 0.15 kt | Yamaguchi et al. (2006) |
| Upper Silesian Basin (Poland) | RECOPOL | 2004 | $CO_2$ | 2w | 0.76 kt | Van Bergen et al. (2006) |
| Black Warrior Basin (USA) | SECARB | 2009 | $CO_2$ | - | 1 kt | Litynski et al. (2008) |
| Central Appalachian Basin (USA) | | 2009 | $CO_2$ | - | 1 kt | Litynski et al. (2008) |
| Illinois Basin (USA) | MGSC | 2008 | $CO_2$ | - | 0.2 kt | Litynski et al. (2008) |
| San Juan Basin (USA) | Coal-Seq | 1995 | $N_2$ | mw | - | Reeves (2004) |
| | | | $CO_2$ | mw | 370 kt | |
| San Juan Basin (USA) | SWP | 2008 | $CO_2$ | - | 75 kt | Litynski et al. (2008) |

vation of this research work, need to be addressed both at laboratory and field test scales to assess the potential of a coal seam for an ECBM operation. They are summarized as follows and will be treated in detail in the next chapters:

- Pure and competitive sorption data on coal of the gases involved in the process are needed, being the former essential for the coal storage capacity estimates and the latter a prerequisite for the description of the displacement dynamics.

- Studies on the coal swelling phenomenon and its consequences on the coal permeability are needed, since they control the gas flow through the coal seam, affecting therefore the overall ECBM operation.

- Finally, ECBM simulation studies aimed at the investigation of the different injection policies represent an essential tool to be used to assess the potential of future ECBM operations.

## 1.6 Structure of the thesis

The objective of this research work is develop experimental and modeling tools that are able to provide a comprehensive coal characterization required to assess the potential of a coal bed for an ECBM operation. Such study has therefore to address the issues just mentioned above, and the thesis is so structured as to follow such a characterization process. Once a coal bed has been identified for a potential ECBM operation, samples are provided to be analyzed in the laboratory. Chapter 2 presents pure gas sorption experiments on coal; by analyzing several coals from different coal mines worldwide the effect of several parameters

## 1.6 Structure of the thesis

such as the temperature, the depth and coal properties on the sorption capacity have been investigated.

Chapter 3 deals with multi-component sorption experiments on coal; the ECBM operation is in fact controlled by an adsorption/desorption process and therefore competitive sorption equilibria involving $CO_2$ and $CH_4$ are required, which are the two main components of any ECBM operation. These measurements involve also nitrogen ($N_2$), which has been successfully used as a co-injectant with $CO_2$, due to the permeability problems caused by the swelling of the coal.

Upon gas sorption coal expands; in order to quantify this phenomenon, a visualization technique has been applied and presented in Chapter 4 allowing to estimate the volumetric behavior of coal upon exposure to an atmosphere of an adsorbing gas such as $CO_2$, $CH_4$ and $N_2$ as well as to the inert helium.

As suggested by the first field tests studies, the swelling phenomenon has dramatic consequences on the coal permeability. In Chapter 5, an experimental technique has been developed and presented to carry out gas injection experiments at conditions similar to those encountered underground: injection of different gases in to coal cores confined by an external pressure is performed. A mathematical model consisting of mass balances accounting for gas flow and sorption, and mechanical constitutive equations for the description of porosity and permeability changes during injection is used to describe the experiment, thus allowing to track the changes in permeability during gas injection.

All the outcomes of the above mentioned experimental studies are the needed information to be included in ECBM simulation studies, aimed at the description and design of ECBM processes. A modeling study has been undertaken and presented in Chapter 6. The coal core model used to describe the pure gas injection experiments has been extended to mixtures, allowing us to study the gas flow process in coal beds and the displacement dynamics during an ECBM operation by investigating the effect of different injection policies (ECBM schemes with different injection composition) on the performance of the ECBM process.

Once the $CO_2$ has been injected, it has to be permanently retained in the geological structure. In Chapter 7, the main mechanisms that prevent the $CO_2$ to leak out of the reservoir are presented together with those controlling its release, with the aim of identifying and quantify their time scale.

Finally, Chapter 8 gives a brief outlook on research which has been recently undertaken and should be pursued in the future.

# Chapter 2

# Sorption on coal: pure gases

## 2.1 Introduction

With increasing interest in $CO_2$ storage in coal seams, more sorption data on coal samples from several mines worldwide become available. This raises a number of questions regarding the reliability of these data and in particular their use for ECBM applications. Some of these issues are highlighted in the following.

First, the gas uptake process in coal has a dual nature, being a combination of adsorption on its surface and penetration (absorption) into its solid matrix. One of the consequences of sorption is that coal changes its volume. Likewise the dual nature of the sorption process, also coal's volumetric behavior (swelling) can be interpreted in two complementary ways. On the one hand, coal expansion may be understood as a consequence of the purely physical adsorption process: adsorption induces a change of the coal specific surface energy, which can be compensated by the elastic energy change associated to the volume change (Jakubov and Mainwaring, 2002; Pan and Connell, 2007). On the other hand, as a glassy, strained, cross-linked macromolecular system, coal undergoes structural changes in the presence of high pressure gas that can be explained only by penetration of the fluid into the coal matrix (Karacan, 2003; Larsen, 2004). Independently of whether one or the other mechanism is responsible for swelling, these volumetric changes need to be taken into account, when interpreting the measured sorption data. Moreover, most conventional techniques used to perform sorption experiments on coal do not allow to separate between the effects of adsorption and absorption; unfortunately this is not always acknowledged in the literature reporting sorption data on coal.

Secondly, to be useful for field applications, the experiments in the laboratory need to reproduce as closely as possible the underground conditions in the coal reservoir. In particular, since the coal seams to be exploited for storage purposes are very deep, it is necessary to perform these experiments at high temperature and pressure. The definition of a standard procedure for measuring sorption isotherms accurately at conditions relevant for ECBM application is therefore of key importance. This can be attained by comparing data obtained on the same samples, but measured in different laboratories using different techniques. Such comparative studies among laboratories are however just beginning. As an example, the U.S. Departement of Energy initiated recently a series of studies on Argonne Premium coal samples, where the most commonly used techniques to measure adsorption isotherms were compared, namely the manometric, volumetric and gravimetric methods (Goodman et al.,

## 2.1 Introduction

2004, 2007). For both dry and moisture-equilibrated samples, the reported $CO_2$ sorption data diverged significantly among the laboratories suggesting that further studies are needed.

In addition, coal seams present a characteristic structure, where layers of coal are separated by thin rock bands (usually shales) (Van Krevelen, 1981). Samples taken at different depths, and therefore belonging to different coal layers, may have different properties, i.e. sorption characteristics (Bromhal et al., 2005; Korre et al., 2007). These spatial variations need to be taken into consideration when performing an assessment study of the $CO_2$ storage potential on the whole coal reservoir.

Finally, coal is a mixture of many kinds of organic and inorganic materials, thus exhibiting a relatively high variability in chemical and physical properties (Van Krevelen, 1981; Mukhopadhyay and Hatcher, 1993). Correlations between such properties and gas sorption are desirable. They could be used when comparing sorption isotherms measured on samples from different coal seams worldwide, and as a guide in choosing the most suitable coal seams for ECBM. With respect to this, the amount of studies conducted is limited (Siemons and Busch, 2007; Day et al., 2008a) and often experiments have been carried out at low pressures (Mastalerz et al., 2004; Ozdemir et al., 2004; Saghafi et al., 2007). In this study, the issues raised above are investigated and results are presented. In particular, nine different coal samples obtained from different locations were investigated in terms of sorption capacity of $CO_2$, $CH_4$ and $N_2$. Sorption isotherms have been obtained at different temperatures, between 33°C and 60°C and up to 200 bar, i.e. in the range of interest for ECBM applications. A Langmuir-like model, which takes into account the effects of both adsorption and absorption, is proposed to describe the obtained excess sorption curves. To assess the reliability of the obtained sorption data, the reproducibility between the isotherms measured in two different laboratories making use of two different gravimetric set-ups and methods was tested. Parameters affecting coal sorption capacity, such as temperature, depth and rank are also discussed.

## 2.2 Experimental

### 2.2.1 Coal characterization

Nine different coal samples obtained from different locations worldwide were investigated; their proximate composition obtained from thermogravimetric analysis (TGA, see Appendix A) is shown in Table 2.1. All the samples are high volatile bituminous coal (vitrinite reflectance, $R_o$, between 0.7 and 0.9%) with the exception of sample A3 of higher rank (medium volatile bituminous, $R_o$=1.34%). Prior to the sorption measurements, the coal samples were crushed and sieved to obtain the desired particle size. Subsequently, they were all dried in an oven at 105°C under vacuum for 1 day, with the exception of samples A1, A2 and A3 for which the drying procedure was carried out at 60°C (see Section 2.4.1). Table 2.1 reports also the sample helium density, as measured by Helium pycnometry (AccuPyc 1330, Micromeritics, Brussels, Belgium). Such measurements have been carried out at 29°C and at a pressure of about 2 bar. With the exception of sample S1 (see Section 2.4.4), the obtained densities lie in the expected range for coals, i.e. between 1.3 and 1.5 g/cm$^3$ (Van Krevelen, 1981; Levine, 1993). The following pure gases obtained from PanGas (Dagmersellen, Switzerland) were used in this study, namely, $CO_2$ and $CH_4$ at purities of 99.995% and $N_2$ and He at purities of 99.999%.

Coal possesses a complex porous structure characterized by a broad pore size distribution, which extends from the nanometer scale (micropores)

## 2.2 Experimental

Figure 2.1: Reflected light microphoto of a polished section of coal sample I2. Scale bar 200 $\mu$m.

up to apertures of several microns, the so-called cleats (Mahajan, 1991; Close, 1993). The presence of the cleats is determinant, since they allow for gas flow through the coal, whereas the micropores contribute to the surface area where gas can adsorb. The characteristic pore structure described above can be best visualized by looking at microphotos of a polished section of coal, as shown in Figure 2.1 for coal sample I2. In particular, the cleats and their structure can be easily recognized in the figure. They form a continuous network, which separate regions where microfractures and pores are present.

### 2.2.2 Experimental methods

High pressure adsorption isotherms were obtained using a Magnetic Suspension Balance (Rubotherm, Germany), whose characteristics and details are extensively described elsewhere (Keller and Staudt, 2005; Ottiger et al., 2006). A typical adsorption experiment consists of the following steps: the high pressure cell containing the powdered coal sample (about 3 g) is evacuated and the weight under vacuum is measured.

Table 2.1: Main properties of the nine coals investigated. [a]Ref. (Nagra, 1989). [b]Ref. (Sakurovs et al., 2007).

| Coal origin | Japan | Italy | | Switzerland | | | Australia | | |
|---|---|---|---|---|---|---|---|---|---|
| Sample | J1 | I1 | I2 | S1[a] | S2[a] | S3[a] | A1[b] | A2 | A3 |
| Moisture (%) | 1.90 | 7.80 | 5.32 | 1.00 | 1.00 | 0.80 | - | 1.63 | 0.39 |
| Volatile Matter (%) | 36.37 | 30.99 | 40.25 | 23.70 | 28.80 | 26.70 | 24.20 | 25.59 | 17.65 |
| Fixed Carbon (%) | 58.85 | 50.09 | 45.72 | 43.50 | 53.50 | 44.20 | 54.20 | 56.12 | 64.16 |
| Ash (%) | 2.88 | 11.12 | 8.71 | 31.80 | 16.70 | 28.30 | 21.60 | 12.66 | 17.80 |
| $R_o$ (%) | 0.78 | 0.74 | 0.70 | - | 0.85 | 0.90 | 0.81 | 0.80 | 1.34 |
| Density (g/cm$^3$) | 1.34 | 1.44 | 1.40 | 1.73 | 1.37 | 1.34 | 1.47 | 1.49 | 1.38 |
| Particle Size ($\mu$m) | 90-180 | <250 | 250-355 | 250-355 | | | 500-1000 | | |

## 2.2 Experimental

Then, the system is filled with helium to obtain the volume of the metal parts and of the coal sample. After evacuating it again, the cell is filled with the gas to be adsorbed, i.e. $CO_2$, $CH_4$ or $N_2$, and the weight is measured at the desired conditions.

For a typical non-swelling adsorbent, e.g. zeolites, this technique allows measurement of the excess mass adsorbed, i.e.

$$n^{ex}(\rho^b, T) = n^a - \rho^b V^a = \frac{M_1(\rho^b, T) - M_1^0 + \rho^b V^0}{M_m m_0^{ads}} \quad (2.1)$$

where $n^{ex}$ is the molar excess adsorbed amount, $n^a$ is the total (absolute) adsorbed amount, $\rho^b$ the molar bulk fluid density and $V^a$ the volume of the adsorbed phase. The right-hand side of Equation (2.1) contains only measurable quantities, that is, the balance signals $M_1(\rho^b, T)$ and $M_1^0$ measured at the desired conditions and under vacuum, the density of the bulk phase $\rho^b$ (measured in-situ through a calibrated titanium sinker) and $V^0$, the sum of the volumes of the metal parts and the initial adsorbent volume. It is worth noting that $V^0$ has been measured with helium at the regeneration temperature of the coal sample (either 60°C or 105°C) as suggested by Malbrunot et al. (1997). $M_m$ and $m_0^{ads}$ are the gas molar mass and the mass of the adsorbent, respectively.

In the case of coal, the uptake of $CO_2$, $CH_4$ and $N_2$ is a combination of adsorption on its surface and penetration (absorption) into its solid matrix, both resulting in coal swelling. As given by the right-hand side of Eq.(2.1), the only truly measurable quantity accounts therefore for the effect of both adsorption and absorption, whose contributions cannot be separated, and is given by (Ottiger et al., 2008a):

$$n^{eas}(\rho^b, T) = n^a + n^s - \rho^b(V^a + \Delta V^s) \quad (2.2)$$

where $n^{eas}$ is the excess sorption and $n^s - \rho^b \Delta V^s$ the absorption term corrected for the buoyancy, with $\Delta V^s$ defined as the difference between the volume of the mixture of coal and imbibed $CO_2$, and the initial sample volume (Rajendran et al., 2005).

## 2.2.3 Absolute sorption isotherms

In this section, a method purely based on experimental observation is presented, in order to obtain the absolute sorption isotherm from the measured excess sorption isotherm. Figure 2.2 shows the $CO_2$ excess sorption isotherm obtained for coal sample I2 at 45°C as a function of the bulk density.

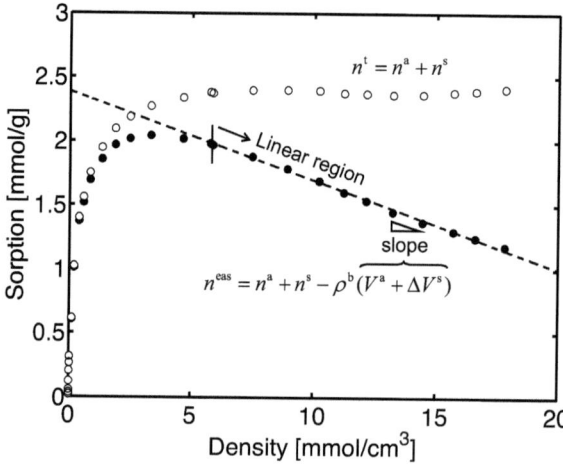

Figure 2.2: $CO_2$ excess sorption amount $n^{eas}$ (●) obtained for coal sample I2 at 45°C as a function of the bulk density. The total uptake $n^t = n^a + n^s$ (○) has been obtained by applying the graphical estimate method (Sudibandriyo et al., 2003b).

## 2.2 Experimental

At high densities, the excess sorption decreases linearly with density, in accordance with several other studies on coal (Fitzgerald et al., 2005; Bae and Bhatia, 2006; Ottiger et al., 2006; Sakurovs et al., 2007; Day et al., 2008a). This behavior has also been observed for various commercial adsorbents, such as activated carbon (Humayun and Tomasko, 2000; Sudibandriyo et al., 2003b) and zeolites (Hocker et al., 2003; Gao et al., 2004), thus indicating that both volume and density of the adsorbed phase become constant, as indicated by Eq.(2.1). An analogous interpretation can be applied to coal, where, since both adsorption and absorption are present, the term becoming constant in Eq.(2.2) is therefore $\hat{V} = (V^a + \Delta V^s)$. In other words, in the linear region of the excess isotherm, where the density of the fluid phase is large, the coal becomes saturated. From the slope of the linear region the quantity $\hat{V}$ can be estimated; in the case of coal sample I2 $\hat{V} = 0.069$ cm$^3$/g. By assuming that the value of $\hat{V}$ is constant over the whole density range (constant volume assumption, Murata et al. (2001)), the total uptake $n^t = n^a + n^s$ can be obtained, as shown in Figure 2.2. It can be seen that the obtained isotherm is of type I (Ruthven, 1984), characterized by a gradual flattening (saturation limit) at large densities. It is worth noting that some studies hint at an irregular behavior in the excess sorption isotherm of $CO_2$, particularly close to its critical density, with deviations from the linearity observed here (Krooss et al., 2002; Toribio et al., 2005; Romanov et al., 2006). It is believed that these were artifacts due to the experimental set-ups and procedures, as demonstrated by the fact that when repeating one of the measurements in another setting the irregularities disappeared (Toribio et al., 2005; Romanov et al., 2006). In this context, the impact of several sources of error on the measured high-pressure adsorption isotherms have been recently discussed in detail (Sakurovs et al., 2008a).

## 2.2.4 Swelling

Uptake and release of gases and liquids are associated with swelling and shrinking of coal, respectively (Larsen, 2004). Volume changes of unstressed coal samples of a given shape upon exposure to a gas at high pressure for several days can be measured using the dilatometric, optical or strain measurement methods. In most cases it has been reported that the extent of swelling increases monotonically with pressure up to a few percents for adsorbing gases, with $CO_2$ swelling coal more than $CH_4$ that swells it more than $N_2$, whereas for helium, a non-adsorbing gas, volume changes are negligible (Harpalani and Chen, 1995; St. George and Barakat, 2001; Day et al., 2008b; Ottiger et al., 2008a; Cui et al., 2007). Moreover, it has been shown that the measured swelling can be effectively described by Langmuir-like curves (Levine, 1996; Palmer and Mansoori, 1998; Shi and Durucan, 2004a; Cui et al., 2007; Pini et al., 2009), e.g.,

$$s = \frac{s^{\max} b_{\mathrm{v}} \rho^{\mathrm{b}}}{1 + b_{\mathrm{v}} \rho^{\mathrm{b}}} \qquad (2.3)$$

with parameters $s^{\max}$ and $b_{\mathrm{v}}$. Isotropic swelling of a coal disc from the same batch as sample I2 has been measured at 45°C with $CO_2$ in our laboratory as reported in a previous study (Ottiger et al., 2008a). The experimental data have been fitted to a Langmuir-like function thus yielding the values $s^{\max} = 0.043$ and $b_{\mathrm{v}} = 1.19$ cm$^3$/mmol (Mazzotti et al., 2009).

## 2.3 Sorption isotherm model

### 2.3.1 Adsorption and absorption

Due to the complexity of the gas uptake process in coal, recent studies have proposed the use of hybrid models, where the two components of the sorption mechanism (adsorption and absorption) are described separately (Milewska-Duda, 1987; Ozdemir et al., 2003; Sakurovs et al., 2007). Two different isotherm equations are considered in order to describe the adsorption on coal, namely the Langmuir and the Dubinin-Radushkevich (DR) isotherms:

$$\text{Langmuir}: \quad n^{\text{a}} = \frac{n_{\text{a}}^{\max} b_{\text{a}} \rho^{\text{b}}}{1 + b_{\text{a}} \rho^{\text{b}}} \quad (2.4\text{a})$$

$$\text{Dubinin} - \text{Radushkevich}: \quad n^{\text{a}} = n_{\text{a}}^{\max} \exp\left\{-D\left[\ln\frac{\rho^{\text{a}}}{\rho^{\text{b}}}\right]^2\right\} \quad (2.4\text{b})$$

where, in the Langmuir equation, $n_{\text{a}}^{\max}$ and $b_{\text{a}}$ are the saturation capacity per unit mass coal and the Langmuir equilibrium constant, respectively. In the Dubinin-Radushkevich equation, $\rho^{\text{a}}$ is the adsorbed phase density, $n_{\text{a}}^{\max}$ is the adsorption capacity as $\rho^{\text{a}} = \rho^{\text{b}}$, and $D$ is a constant related to the affinity of the sorbent for the gas. Since many experimental isotherms are of type I (Ruthven, 1984), the Langmuir model allows for a reasonably good fit by proper choice of the parameters $n_{\text{a}}^{\max}$ and $b_{\text{a}}$. Probably this is the reason why this model has been widely applied in most reservoir simulators used to predict the dynamics of ECBM processes (Shi and Durucan, 2004b; Bromhal et al., 2005; Smith et al., 2005; Shi and Durucan, 2006; Bustin et al., 2008; Shi et al., 2008). On the other hand, the DR model has been particularly successful in the description of subcritical adsorption on microporous solids such as activated carbon

(Dubinin and Stoeckli, 1980; Hutson and Yang, 1997) and coal (Ozdemir et al., 2003) and has been recently extended to supercritical conditions (Sakurovs et al., 2007).

A Langmuir-like model and the Henry's law are considered for absorption to be combined with the adsorption isotherm above:

$$\text{Langmuir}: \quad n^s = \frac{n_s^{\max} b_s \rho^b}{1 + b_s \rho^b} \tag{2.5a}$$

$$\text{Henry}: \quad n^s = k \rho^b \tag{2.5b}$$

Both Langmuir-like isotherm (see Section 2.2.4) and the linear equation (Ozdemir et al., 2003; Sakurovs et al., 2007) have been used earlier. The total gas uptake is then given by the sum of the contributions of adsorption and absorption, i.e. $n^t = n^a + n^s$.

## 2.3.2 Proposed model for sorption

In chromatography, a so-called Bi-Langmuir model is usually applied to describe adsorption on a surface exhibiting two types of adsorption sites, each subject to an independent Langmuir isotherm (Fornstedt et al., 1996). For analogy, the same equation is proposed here to describe the combination of adsorption, Eq.(2.4a), and absorption, Eq.(2.5a), respectively. Since both equations have identical shape, two constraints are imposed to be able to distinguish between the two uptake mechanisms. First we assume that the swelling and the absorption isotherms have identical shape, and therefore that the parameter $b_s$ is equal to the experimentally obtained $b_V$ (Section 2.2.4). Secondly the combined isotherm has to be consistent with the observed behavior at very low density, i.e.,

## 2.3 Sorption isotherm model

$$\frac{n^t}{\rho^b} = (n_a^{max}b_a + n_s^{max}b_s) = H \quad \text{for} \quad \rho^b \to 0 \qquad (2.6)$$

with $H$ being the measured Henry's constant. For coal sample I2, fitting the measured sorption data with a straight line at very low densities allowed to obtain a value of $H$=10.66 cm$^3$/g. Figure 2.3 shows the total $CO_2$ uptake $n^t$, as obtained by following the procedure explained in Section 2.2.3, as a function of the bulk density. Along with the measured values is shown the model prediction (black solid line), as obtained by fitting the Bi-Langmuir model to the experimental data, together with the contribution of adsorption and absorption (dashed lines). It can be seen that the agreement between experiment and model is rather satisfactory, and that absorption account for about 40% of the total $CO_2$ uptake. Values of the fitted parameters are summarized in Table 2.2.
The Bi-Langmuir equation can be further simplified by assuming that $b_a = b_s = b$, thus resulting in the following Langmuir-like equation:

$$n^t = \frac{(n_a^{max} + n_s^{max})b\rho^b}{1 + b\rho^b} = \frac{n^{max}b\rho^b}{1 + b\rho^b} \qquad (2.7)$$

having only two fitting parameters, $n^{max}$ and $b$, which now lump together the contribution of adsorption and absorption. Eq.(2.7) has also been fitted to the experimental sorption data resulting in the gray solid line shown in Figure 2.3. Values of the fitted parameters are reported in Table 2.2. Also in this case the model prediction is satisfactory, as demonstrated by the fact that the difference between the two fitting functions is barely visible, which suggests that also a simple Lanmguir-like equation is able to describe the gas uptake process in coal. It is worth pointing out that, although in this case no distinction between the two components of the sorption mechanism can be made, the use of Eq.(2.7) does not imply that one or the other component is neglected. Moreover, the use of Eq.(2.7) is very useful from a practical point of view. First, it does not require a detailed knowledge of the sorption isotherm in the low density range to obtain the Henry's constant. Secondly, it does not require an independent measure of swelling: in this respect the amount of studies at condition relevant to ECBM is in fact very limited, when compared to the available literature on high-pressure sorption on coal.

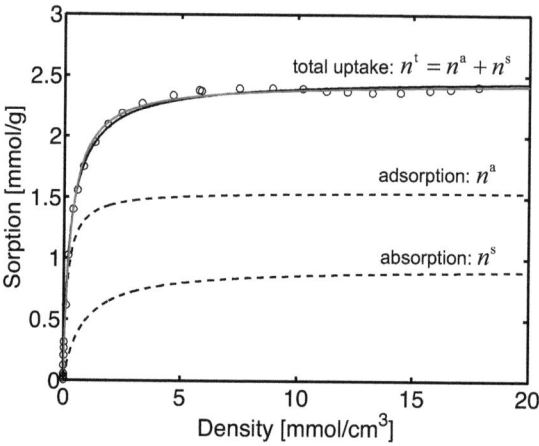

Figure 2.3: $CO_2$ sorption isotherm $n^t$ (○) obtained for coal sample I2 at 45°C as a function of the bulk density. Lines represent model results from two different isotherm equations: Bi-Langmuir model (black solid line) with corresponding component contributions (dashed lines), and Langmuir-like model, Eq.(2.7) (gray solid line).

## 2.3 Sorption isotherm model

Table 2.2: Model parameters for $CO_2$ sorption on coal sample I2 at 45°C from two different isotherm equations: Bi-Langmuir model and Langmuir-like model, Eq.(2.7).

| Bi-Langmuir | | Langmuir | |
|---|---|---|---|
| $n_a^{max}$ (mmol/g) | 1.55 | $n^{max}$ (mmol/g) | 2.45 |
| $b_a$ (cm$^3$/mmol) | 6.14 | $b$ (cm$^3$/mmol) | 3.35 |
| $n_s^{max}$ (mmol/g) | 0.93 | | |
| $b_s$ (cm$^3$/mmol) | 1.19 | | |
| $\hat{R}$ (%) | 1.52 | $\hat{R}$ (%) | 1.61 |

### 2.3.3 Comparison with literature

As mentioned above, other models can be used to describe the experimentally obtained gas sorption isotherms on coal. In particular, the following combination of the DR equation (adsorption term) with a term proportional to gas density following Henry's law (absorption term), has been investigated in previous studies (Ozdemir et al., 2003; Sakurovs et al., 2007):

$$n^{eas}(\rho^b, T) = n_a^{max} \exp\left\{-D\left[\ln\frac{\rho^a}{\rho^b}\right]^2\right\} \left(1 - \frac{\rho^b}{\rho^a}\right) + k\rho^b \qquad (2.8)$$

with fitting parameters $n^{max}$, $D$ and $k$. The adsorbed phase density takes the value of 22.7 mmol/cm$^3$ (Sakurovs et al., 2007).

We have applied this model to the excess data obtained for coal sample I2 and we have compared the results with those obtained using the model proposed in this study. To do this, the Langmuir equation, i.e. Eq.(2.7), is recast in its excess form as

$$n^{eas}(\rho^b, T) = \frac{n^{max} b \rho^b}{1 + b\rho^b} - \rho^b V \qquad (2.9)$$

with fitting parameters $n^{max}$, $b$ and $V = V^a + \Delta V^s$. Figure 2.4 shows the $CO_2$ excess sorption data obtained for coal sample I2 together with the prediction from the two isotherm models. It can be seen that both models are able to reproduce the excess data in a reasonably good way, but that the DR equation combined with the Henry's law fails to predict the total $CO_2$ uptake.

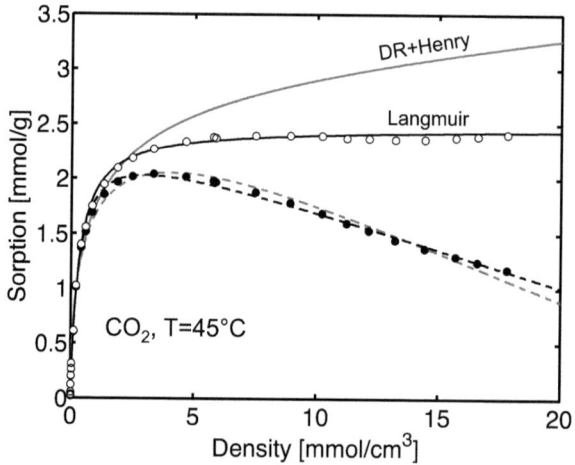

Figure 2.4: $CO_2$ excess sorption isotherm $n^{eas}$ (●) obtained for coal sample I2 at 45°C as a function of the bulk density and corresponding total uptake $n^t$ (○). Lines represent model results from two different isotherm equations: Langmuir-like model, Eq.(2.9) (black lines) and DR equation combined with Henry's law, Eq.(2.8) (gray lines).

The fitted parameters for both methods are reported in Table 2.3. Interestingly, the fitted value for $V$ (0.071 cm$^3$/g) in the Langmuir-like model is very close to the one obtained in Section 2.2.3 with the graphical method, i.e. $\hat{V} = 0.069$ cm$^3$/g. The main problem in the use of Eq.(2.8) is that it neglects the swelling effect, i.e. $\Delta V^s$, and it is therefore

## 2.3 Sorption isotherm model

inconsistent in the interpretation of the experimental data. Although this equation is able to catch fairly well the excess sorption behavior of this specific coal, the overestimation of the total $CO_2$ uptake makes it useless for any gas storage capacity estimation for ECBM operations, where a good prediction of the total amount of $CO_2$ potentially stored is needed.

Table 2.3: Model parameters for $CO_2$ sorption on coal sample I2 at 45°C from two different isotherm equations: Langmuir-like model, Eq.(2.9) and DR equation combined with Henry's law, Eq.(2.8).

| Langmuir | | DR-Henry | |
|---|---|---|---|
| $n^{\text{max}}$ (mmol/g) | 2.47 | $n^{\text{max}}$ (mmol/g) | 2.70 |
| $b$ (cm$^3$/mmol) | 3.26 | $D$ | 0.045 |
| $V$ (cm$^3$/g) | 0.071 | $k$ (cm$^3$/g) | 0.028 |
| $\hat{R}$ (%) | 1.58 | $\hat{R}$ (%) | 2.02 |

In the light of these results, only Eq.(2.9) will be considered in the following sections. In particular, for all the coals used in this study, the experimentally obtained excess sorption has been fitted and values for the parameters $n^{\text{max}}$, $b$ and $V$ were determined by minimizing the root mean squared error:

$$R = \frac{1}{N}\sqrt{\sum_{j=1}^{N}\left(n^{\text{eas}}_{\text{exp},j} - n^{\text{eas}}_{\text{mod},j}\right)^2} \qquad (2.10)$$

where $N$ is the number of experimental data points, and the subscripts exp and mod refer to the variables obtained from the experiments and from the model, respectively. The values of the root mean squared error $R$ are normalized by the corresponding maximum adsorption capacity $n^{\text{max}}$, i.e. $\hat{R} = R/n^{\text{max}}$, thus allowing to compare the quality of the fitting for coals with different adsorption capacities. The parameter $V$

accounts for the volume changes caused by adsorption and absorption, i.e. $V = V^{\mathrm{a}} + \Delta V^{\mathrm{s}}$, as given by Eq.(2.2). Since for $CH_4$ and $N_2$ the range of density investigated in this study doesn't cover the linear descending part, a value of $V$ is imposed in the case of these two gases. In particular, this value has been estimated by assuming proportionality with respect to $CO_2$ i.e.,

$$V_i = \frac{n_{i,M}^{\mathrm{eas}}}{n_{CO_2,M}^{\mathrm{eas}}} V_{CO_2} \qquad (2.11)$$

where $n_{i,M}^{\mathrm{eas}}$ is the maximum measured excess sorption amount of component $i$ ($= CH_4$ or $N_2$).

## 2.4 Results

### 2.4.1 Comparison among different laboratories

Single component sorption isotherms of $CO_2$, $CH_4$ and $N_2$ at 55°C on the same Australian coal samples A1 and A2 have been measured in two different laboratories, namely at CSIRO (Newcastle, Australia) as reported earlier (Sakurovs et al., 2007) and in our lab at ETH Zurich (Zurich, Switzerland) using the equipment and procedure described above. Each sample was divided in two parts, one of which was analyzed at CSIRO, and one was sent for analysis to ETH Zurich. Sorption data at CSIRO and ETH Zurich have thus been obtained on initially identical coal. Once received in our lab, care was taken to follow the same procedure as at CSIRO when preparing the sample for the experiments. The sample was kept in a plastic bottle (as received) and was vacuum dried overnight at 60°C prior to the sorption measurements. The techniques to obtain the

## 2.4 Results

sorption isotherms are different in the two labs, but they both rely on a gravimetric approach. In particular, in the set-up used at ETH Zurich only the coal sample basket, which "swims" in the high pressure gas, is weighted and the density is measured with a calibrated sinker (Ottiger et al., 2006), whereas at CSIRO the whole system (coal + cell + fluid) is weighed (Sakurovs et al., 2007) and the density is measured in an additional empty cell. Moreover, for the experiments reported below, different amounts of sample have been used: 200 g at CSIRO and 3 g at ETH Zurich. Results of these experiments are shown in Figure 2.5, where the molar excess sorption $n^{\text{eas}}$ of $CO_2$, $CH_4$ and $N_2$ at 55°C is plotted as a function of the molar bulk density. It can be seen that for both samples and for all gases the agreement between the two laboratories over the whole range of densities (which for all gases corresponds to pressures up to 200 bar) is rather satisfactory. Together with the experimental points are shown the Langmuir equations fitted for each gas to both sets of data simultaneously. It can be seen that the Langmuir model is able to reproduce well the experimental adsorption data. The values of the fitted parameters are reported in Table 2.4 together with the normalized root mean square error $\hat{R}$.

### 2.4.2 Comparison of different coals

The experimental results shown in Figure 2.6, where the molar excess sorption $n^{\text{eas}}$ is plotted against the bulk density $\rho^{\text{b}}$, refer to the sorption of $CO_2$, $CH_4$ and $N_2$ on all coal samples considered in this study measured at 45°C and up to 190 bar. For all coals examined the experimental data behave similarly and follow the typical behavior of excess adsorption isotherms. In the case of $CO_2$, the excess sorption increases with the bulk density to reach a maximum, located at around $\rho^{\text{b}} =$

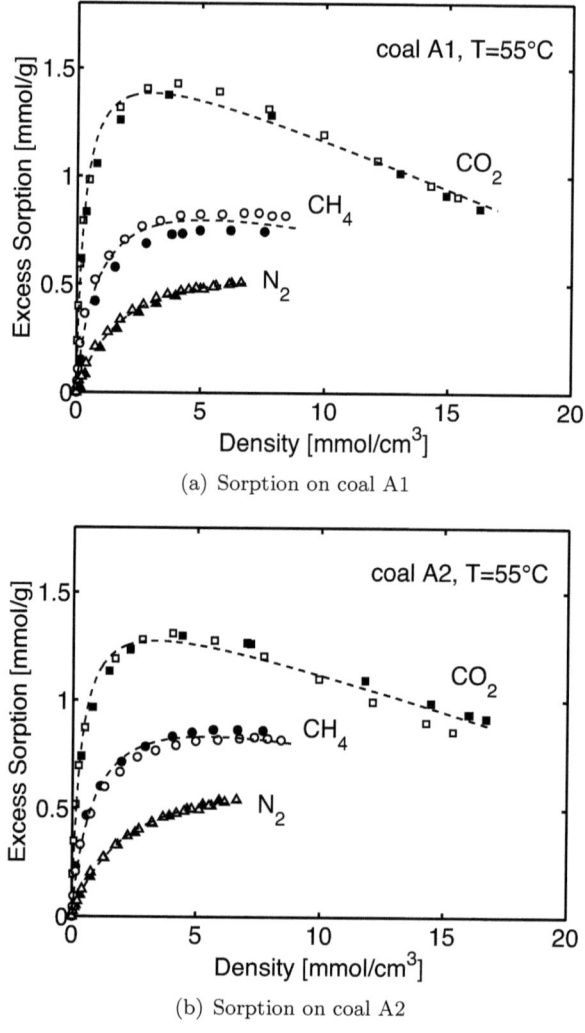

(a) Sorption on coal A1

(b) Sorption on coal A2

Figure 2.5: Comparison among different laboratories. $CO_2$, $CH_4$ and $N_2$ molar excess sorption $n^{eas}$ on coal samples (a) A1 and (b) A2 at 55°C as a function of the bulk density measured at CSIRO (Newcastle, Australia) (open symbols) (Sakurovs et al., 2007) and in our lab (closed symbols).

## 2.4 Results

Table 2.4: Langmuir model parameters for $CO_2$, $CH_4$ and $N_2$ sorption on coal samples A1 and A2 at 55°C.

|  | Sample | A1 | A2 |
|---|---|---|---|
| $CO_2$ | $n^{max}$ (mmol/g) | 1.68 | 1.54 |
|  | $b$ (cm$^3$/mmol) | 3.30 | 3.06 |
|  | $V$ (cm$^3$/g) | 0.047 | 0.037 |
|  | $\hat{R}$ (%) | 3.35 | 3.27 |
| $CH_4$ | $n^{max}$ (mmol/g) | 1.09 | 1.13 |
|  | $b$ (cm$^3$/mmol) | 1.13 | 1.14 |
|  | $V$ (cm$^3$/g) | 0.026 | 0.025 |
|  | $\hat{R}$ (%) | 4.08 | 2.67 |
| $N_2$ | $n^{max}$ (mmol/g) | 0.84 | 0.89 |
|  | $b$ (cm$^3$/mmol) | 0.42 | 0.38 |
|  | $V$ (cm$^3$/g) | 0.017 | 0.016 |
|  | $\hat{R}$ (%) | 1.55 | 0.80 |

3.9 mmol/cm$^3$, and decreases linearly by further increasing the density. In the case of $CH_4$, the isotherms exhibit also a maximum, though less pronounced, whereas for $N_2$ this is not visible, being the measuring conditions far above the critical temperature. It is well known that coal adsorbs $CO_2$ more than $CH_4$, and $CH_4$ more than $N_2$ (Fitzgerald et al., 2005; Shimada et al., 2005; Bae and Bhatia, 2006; Sakurovs et al., 2007). This behavior, which has also been observed for all the coals tested here, is a prerequisite for a successful ECBM operation. The theoretical effectiveness of the $CH_4$ displacement by the injected $CO_2$ during the ECBM operation predicted by the pure sorption isotherms need however to be confirmed by multicomponent sorption measurements, where the different gases compete simultaneously for sorption on coal. Binary and ternary mixture sorption measurements on sample I2 performed at a temperature of 45°C and up to 180 bar for 11 different gas mixtures of $CO_2$, $CH_4$ and $N_2$ confirmed the expected behavior and are reported in details elsewhere (Ottiger et al., 2008a,b).

Together with the experimental points, in Figure 2.6 are shown the fitted Langmuir curves, i.e. Eq.(2.9), whose parameters are reported in Table 2.5 together with the normalized root mean squared error ($\hat{R}$). It can be seen that there is a good agreement between experiments and fitted curves: in most cases the difference is in fact smaller than 2%. This result suggests that the Langmuir-like model is a valuable option for the description of gas sorption on coal at these conditions. In the Langmuir equation, the saturation capacity per unit mass of coal, $n^{\max}$, the Langmuir equilibrium constant, $b$ and the term accounting for the volume changes caused by adsorption and absorption, $V = V^a + \Delta V^s$ have been fitted to the experimental data. It is worth highlighting that in the case of $CO_2$ the fitted values of $V$ are very close to the values obtained by applying the graphical estimate method to the experimental data ($\hat{V}$), as already observed in Section 2.2.3. From the fitted values, the maximum value of the density of the adsorbed and absorbed phase can be estimated as $\rho^{\max} = n^{\max}/V$. In the case of $CO_2$, with the exception of coal I1 and A2, the obtained values belong to a rather narrow range, i.e. 35.41 ± 0.72 mmol/cm$^3$. For the sake of comparison, an adsorbed phase density of approximately 22.73 mmol/cm$^3$ has been reported for $CO_2$ adsorption on activated carbon (Humayun and Tomasko, 2000; Sudibandriyo et al., 2003b), whereas for a 13X zeolite a higher value of about 38.41 mmol/cm$^3$ has been measured (Gao et al., 2004).

It can be seen that there is a significant difference in terms of maximum sorption capacity, $n^{\max}$, among the different coals. Coals showing a large $CO_2$ sorption capacity are particularly suitable for an ECBM operation finalized at $CO_2$ storage. Figure 2.7 shows the total $CO_2$ uptake $n^t$, as obtained from the graphical estimate method (Section 2.2.3), as a function of the bulk density, together with the prediction of the Langmuir

## 2.4 Results

Table 2.5: Langmuir model parameters for $CO_2$, $CH_4$ and $N_2$ sorption on different coal samples at 45°C.

|  | Sample | J1 | I1 | I2 | S1 | S2 | S3 | A1 | A2 | A3 |
|---|---|---|---|---|---|---|---|---|---|---|
| $CO_2$ | $n^{max}$ (mmol/g) | 1.85 | 3.19 | 2.47 | 0.75 | 1.31 | 1.15 | 1.75 | 1.66 | 1.23 |
|  | $b$ (cm$^3$/mmol) | 2.47 | 1.71 | 3.26 | 1.97 | 2.70 | 2.44 | 3.34 | 3.00 | 3.34 |
|  | $\rho^{max}$ (mmol/cm$^3$) | 36.27 | 30.67 | 34.79 | 10.56 | 36.39 | 34.85 | 35.00 | 42.56 | 35.15 |
|  | $V$ (cm$^3$/g) | 0.051 | 0.104 | 0.071 | 0.071 | 0.036 | 0.033 | 0.050 | 0.039 | 0.035 |
|  | $\hat{V}$ (cm$^3$/g) | 0.046 | 0.093 | 0.069 | 0.068 | 0.033 | 0.029 | 0.055 | 0.043 | 0.034 |
|  | $\hat{R}$ (%) | 1.68 | 0.53 | 1.58 | 1.27 | 1.13 | 1.15 | 2.06 | 2.11 | 1.37 |
| $CH_4$ | $n^{max}$ (mmol/g) |  | 1.77 | 1.63 |  | 0.91 | 0.82 | 1.12 | 1.20 | 0.94 |
|  | $b$ (cm$^3$/mmol) |  | 0.99 | 1.55 |  | 1.48 | 1.25 | 1.10 | 1.16 | 1.80 |
|  | $\rho^{max}$ (mmol/cm$^3$) |  | 33.4 | 37.91 |  | 37.92 | 37.27 | 40.00 | 48.00 | 37.60 |
|  | $V$ (cm$^3$/g) |  | 0.053 | 0.043 |  | 0.024 | 0.022 | 0.028 | 0.025 | 0.025 |
|  | $\hat{R}$ (%) |  | 1.20 | 2.79 |  | 2.00 | 1.71 | 1.05 | 2.00 | 3.20 |
| $N_2$ | $n^{max}$ (mmol/g) |  | 1.28 | 1.00 |  | 0.67 | 0.66 | 0.94 | 0.92 | 0.72 |
|  | $b$ (cm$^3$/mmol) |  | 0.45 | 0.51 |  | 0.50 | 0.38 | 0.38 | 0.40 | 0.58 |
|  | $\rho^{max}$ (mmol/cm$^3$) |  | 38.79 | 45.45 |  | 44.67 | 47.14 | 49.47 | 57.5 | 45.00 |
|  | $V$ (cm$^3$/g) |  | 0.033 | 0.022 |  | 0.015 | 0.014 | 0.019 | 0.016 | 0.016 |
|  | $\hat{R}$ (%) |  | 0.83 | 1.50 |  | 2.16 | 1.36 | 0.65 | 0.89 | 1.35 |

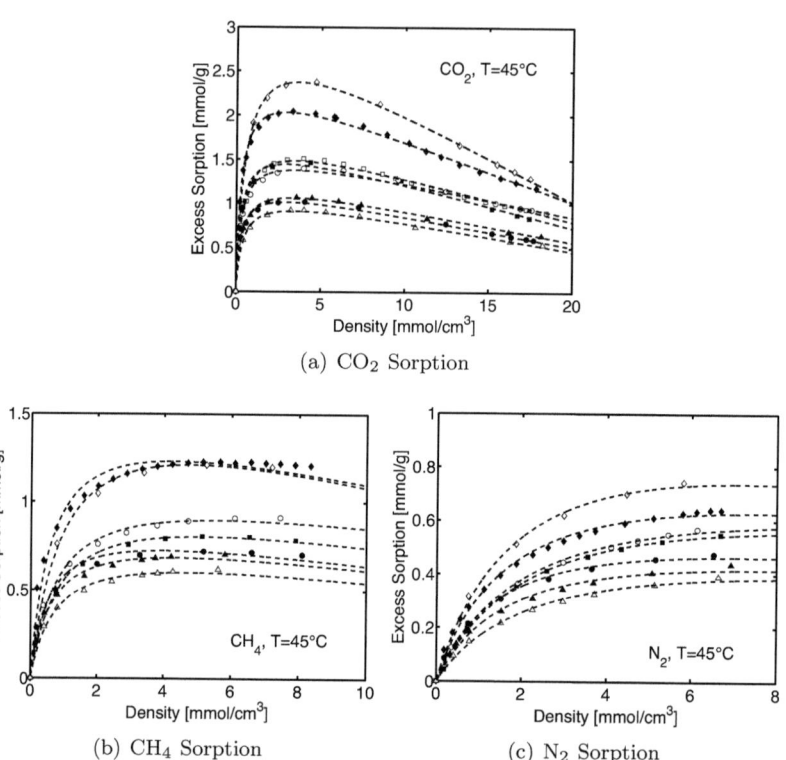

(a) CO$_2$ Sorption

(b) CH$_4$ Sorption

(c) N$_2$ Sorption

Figure 2.6: High pressure pure sorption isotherms on coal. CO$_2$, CH$_4$ and N$_2$ molar excess sorption $n^{eas}$ as a function of the bulk density $\rho^b$ on eight coal samples measured at 45°C. Symbols are experimental points, whereas lines are fitted Langmuir curves. Symbols: I1 ($\Diamond$), I2 ($\blacklozenge$), J1 ($\square$), A1 ($\blacksquare$), A2 ($\bigcirc$), A3 ($\bullet$), S2 ($\triangle$), S3 ($\blacktriangle$).

## 2.4 Results

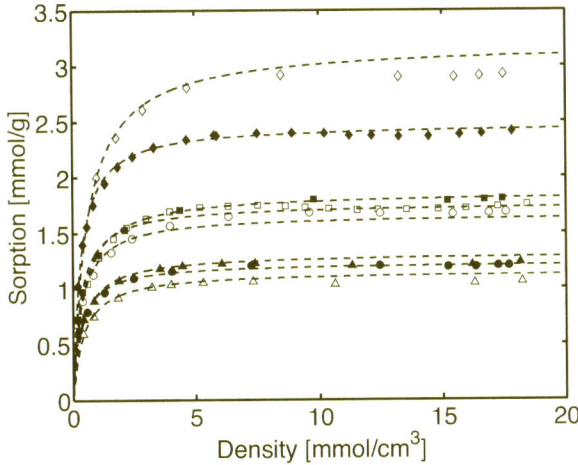

Figure 2.7: Total $CO_2$ uptake $n^t$ as a function of the bulk density $\rho^b$ on eight coal samples measured at 45°C. Symbols are experimental points, whereas lines are fitted Langmuir curves. Symbols: I1 ($\Diamond$), I2 ($\blacklozenge$), J1 ($\Box$), A1 ($\blacksquare$), A2 ($\bigcirc$), A3 ($\bullet$), S2 ($\triangle$), S3 ($\blacktriangle$).

equations, which have been previously fitted to the excess sorption data. Again it can be seen that the agreement between experiments and estimated curves is satisfactory. Moreover, for $CO_2$ the observed maximum adsorption capacity per unit mass of dry coal ranges between 5% and 14% weight. As mentioned in Section 2.2 the regeneration procedure of the coal samples consists in drying them under vacuum for 24 h. For the samples dried at 105°C, we can safely assume that there is no water left on the coal. We believe that this is also true for those dried at 60°C as the repeated charging of the gases in the measuring cell provides additional drying of the sample.

Finally, two comments are worth making with respect to samples comparison. First, as mentioned in Section 2.2 the regeneration procedure of the coal samples consists in drying them under vacuum for 24 h. For the samples dried at 105°C, we can safely assume that there is no water left on the coal. We believe that this is also true for those dried at 60°C as the repeated charging of the gases in the measuring cell provides additional drying of the sample. Secondly, different particle sizes have been used in the measurements, either because the sample was delivered already as a powder (Sample I1) or for the sake of comparison with other studies (Sample J1, A1, A2 and A3). Due to grindability differences, different particle size fractions may include different maceral or different amounts of macerals and therefore they may not be comparable in terms of sorption capacity. However, to our knowledge the studies dealing with this issue do not report specific trends between particle size and gas sorption on coal and the range of particle size investigated is often broader than the one in the present work (Busch et al., 2004; Gruszkiewicz et al., 2009). The most important effect observed was related to the kinetic of the sorption mechanism, which is not important for the static experiments presented in this work.

## 2.4 Results

### 2.4.3 Effect of temperature

For coal sample I1, the sorption of $CO_2$, $CH_4$ and $N_2$ has been measured at three different temperatures, namely 33, 45 and 60°C. Assuming a geothermal gradient of 25°C/km from 15°C at the surface, these temperatures cover the range of depth where $CO_2$ storage in coal seams is considered to be feasible, i.e between 750 and 2000 m. The lower bound corresponds to the depth at which temperature and pressure make it possible to store $CO_2$ as a supercritical fluid, and therefore with a much larger density compared to its gaseous form. The upper bound is given as a limit, where the high drilling costs and the low coal permeability (large overburden stress) would make the operation economically unprofitable. It is worth noting that 33°C is a temperature which is close to the critical temperature of $CO_2$, i.e. 31.0°C. At these conditions the phase behavior of gases like $CO_2$ is strongly dependent on temperature, making the measurement of gas adsorption rather challenging. The magnetic suspension balance used in this work has been the subject of a detailed study, which was carried out to improve its precision and reliability in measuring adsorption especially at conditions close to the critical point (Pini et al., 2006).

In Figure 2.8 the molar excess sorption $n^{eas}$ is plotted against the bulk density $\rho^b$ for $CO_2$, $CH_4$ and $N_2$ at 33, 45 and 60°C. The experimental data follow the typical behavior of excess adsorption isotherms, and over the whole range of density the excess sorption grows with decreasing temperature. It is worth noting that for $CO_2$ and $CH_4$ the maximum of the isotherm moves to higher densities with increasing temperature, in accordance with other studies (Ustinov et al., 2002; Bae and Bhatia, 2006). Together with the experimental points are shown the fitted Langmuir model curves (dashed lines) and the corresponding absolute isotherms (solid lines). The obtained fitted parameters are reported in

Table 2.6: Langmuir model parameters for $CO_2$, $CH_4$ and $N_2$ sorption on coal samples I1 at three temperatures (33, 45 and 60°C).

|  | Temperature | 33°C | 45°C | 60°C |
|---|---|---|---|---|
| $CO_2$ | $n^{max}$ (mmol/g) | 3.44 | 3.19 | 2.94 |
|  | $b$ (cm$^3$/mmol) | 1.97 | 1.71 | 1.44 |
|  | $\rho^{max}$ (mmol/cm$^3$) | 30.71 | 30.67 | 30.62 |
|  | $V$ (cm$^3$/g) | 0.112 | 0.104 | 0.096 |
|  | $\hat{R}$ (%) | 1.06 | 0.53 | 0.51 |
| $CH_4$ | $n^{max}$ (mmol/g) | 1.81 | 1.77 | 1.67 |
|  | $b$ (cm$^3$/mmol) | 1.21 | 0.99 | 0.87 |
|  | $\rho^{max}$ (mmol/cm$^3$) | 32.32 | 33.40 | 33.40 |
|  | $V$ (cm$^3$/g) | 0.056 | 0.053 | 0.050 |
|  | $\hat{R}$ (%) | 1.07 | 1.20 | 1.03 |
| $N_2$ | $n^{max}$ (mmol/g) | 1.26 | 1.28 | 1.20 |
|  | $b$ (cm$^3$/mmol) | 0.55 | 0.45 | 0.39 |
|  | $\rho^{max}$ (mmol/cm$^3$) | 38.18 | 38.78 | 40.00 |
|  | $V$ (cm$^3$/g) | 0.033 | 0.033 | 0.030 |
|  | $\hat{R}$ (%) | 0.80 | 0.83 | 0.70 |

Table 2.6 together with the normalized root mean squared error, $\hat{R}$. Also in this case, it can be seen that the agreement between experiments and model is good: in all cases, the difference between model and experiments is well below 1.5%. It can be seen that the parameter $V$ decreases with increasing temperature; however, when the corresponding maximum density is calculated, i.e. $\rho^{max} = n^{max}/V$, no clear dependence on temperature has been observed: for all gases its value varies only slightly among the different temperatures. For the Langmuir constant $b$, the expected behavior has been found: for all three gases, it decreases with increasing temperature, because of the exothermic nature of the sorption process (Yang, 1997).

The obtained values for the maximum sorption capacity $n^{max}$ decrease with increasing temperature for all three gases, in agreement with other studies on gas sorption on coal (Bae and Bhatia, 2006; Sakurovs et al.,

## 2.4 Results

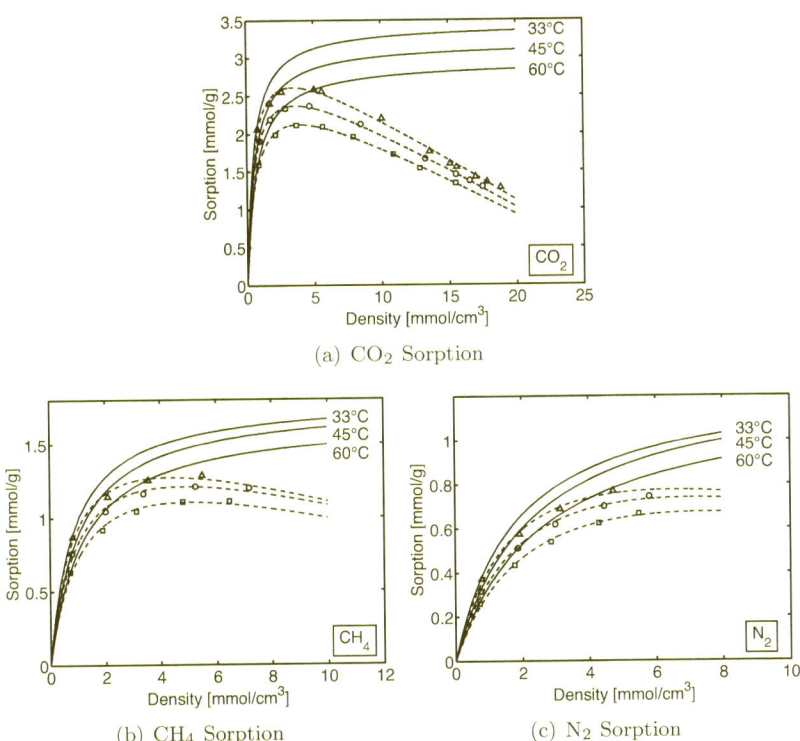

(a) CO₂ Sorption

(b) CH₄ Sorption

(c) N₂ Sorption

Figure 2.8: $CO_2$, $CH_4$ and $N_2$ molar excess sorption $n^{eas}$ on coal sample I1 measured at three different temperatures, namely 33 (△), 45 (○) and 60°C (□). Symbols are experimental points, whereas lines are model results: fitted excess Langmuir curves (dashed lines) and their corresponding absolute isotherms (solid lines).

2008b). Obviously these results are in contrast with the monolayer coverage mechanism assumed by the original Langmuir model, where the maximum adsorption capacity is independent of temperature. However, it should be again highlighted that, although the model used here has a Langmuir-like shape, it is used to describe a different mechanism (adsorption and absorption) from the one originally proposed (only adsorption). As a consequence, the maximal sorption capacity has been let to be dependent on temperature in order to obtain the best fit to the experimental data. An accurate description of the experimental data is in fact the information needed for practical application such as reservoir modeling of ECBM processes. Moreover, most of them deal with Langmuir equations to describe gas sorption on coal (Shi and Durucan, 2004b; Bromhal et al., 2005; Smith et al., 2005; Shi and Durucan, 2006; Bustin et al., 2008; Shi et al., 2008), and through the parameters $b$ and $n^{\max}$ obtained in this study the sorption isotherm is fully described, as discussed in Section 2.3.

### 2.4.4 Spatial variation of the sorption behavior

The coal cores S1, S2 and S3 from Switzerland (Weiach, ZH) used in this study were obtained from a well drilled to a total depth of about 2000 m, which crossed several coal seams. Cores were taken at depths of 1586 m (S2), 1701 m (S3) and 1743 m (S1) for analysis in the laboratory. Samples S2 and S3 can be considered as representative of the corresponding coal layers, whereas S1 is a silty and therefore poorer coal sample. As explained below, this last sample can be considered representative of the transitions rock bands between two thin layers of almost pure coal. Beside the proximate analysis that shows very high ash content, this conclusion is supported by the measured bulk rock density, which is

## 2.4 Results

very high compared to all the other coal samples (see Table 2.1).

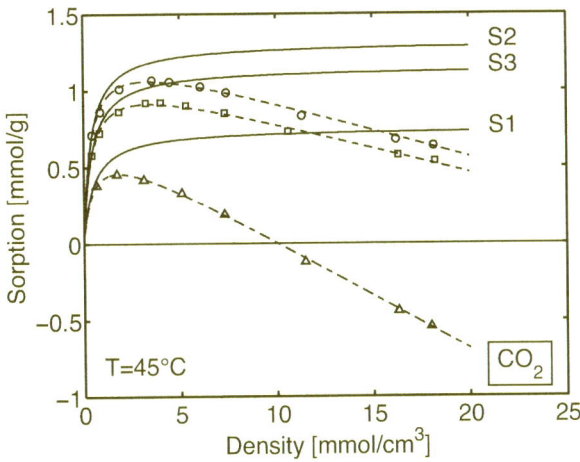

Figure 2.9: $CO_2$ molar excess sorption $n^{eas}$ as a function of the bulk density $\rho^b$ measured at 45°C on coal samples collected at different depths: S1 (△) at 1743 m, S2 (□) at 1586 m and S3 (○) at 1701 m. Symbols are experimental points, whereas lines are model results: fitted excess Langmuir curves (dashed lines) and their corresponding absolute isotherms (solid lines).

Figure 2.9 shows the measured $CO_2$ molar excess sorption $n^{eas}$ as a function of the bulk density $\rho^b$ at 45°C for the three samples S1, S2 and S3. The experimental points have been fitted with the Langmuir model (Eq.(2.9), dashed lines) and they are shown in Figure 2.9 together with the corresponding isotherms in terms of total sorption $n^t$ (solid lines). Qualitatively the isotherms behave in a similar manner, i.e. like a typical excess adsorption isotherm exhibiting a maximum followed by a linear descending part, but they exhibit different $CO_2$ sorption capacities. Moreover, samples S2 and S3 fall down approximately parallel to each other, whereas for sample S1 the decrease is steeper thus leading to a negative excess sorption at high bulk densities. Negative values for the excess sorption are attained when the term $\rho^b V$ is larger than $n^t$, or, equivalently, when the bulk phase density $\rho^b$ exceeds the density of the adsorbed and absorbed phase, $\rho = n^t/V$, as given by Eq.(2.9). This phenomenon doesn't have to be necessarily related to the swelling of coal, since it has been observed also for non-swelling adsorbents (Sircar, 1999). For sample S1 this happens at lower bulk densities compared to samples S2 and S3, where no negative excess is observed in the broad range of pressures studied. This observation is supported by the maximum value of $\rho^{max}$ obtained for S1 (10.6 mmol/cm$^3$), which is much lower compared to samples S2 and S3 (36.3 and 34.8 mmol/cm$^3$, respectively). It is worth noting that the low value of $\rho^{max}$ is not a fitting artifact, but it is directly related to an experimental observation (the parameter $\hat{V}$, i.e. the slope of the linear descending part of the sorption isotherm) and to the low sorption capacity observed for this sample. It is believed that the reason for this different behavior is that, not being pure coal, sample S1 possesses completely different characteristics; the impurities, e.g. shales, reduce its sorption capacity. For samples S2 and S3 maximum sorption capacities between 5 and 6% weight per unit mass coal are obtained, whereas sample S1 exhibits a lower value (3% weight per unit mass coal).

## 2.4.5 Effect of rank

Adsorption studies aim at establishing correlations between sorption capacity and coal properties, in order to guide the choice of the coal seams which are suitable for ECBM. On the one hand, coal exhibits a relatively high variability in chemical and physical properties, being a mixture of many kinds of organic and inorganic materials, the so-called macerals

## 2.4 Results

(Mukhopadhyay and Hatcher, 1993). On the other hand, it is known that many of these properties vary quite regularly with its rank, i.e. the stage the coal has reached in the coalification process, often referred to as its thermal maturity. As an example, the vitrinite reflectance $R_o$ increases with rank and it is the most used and accepted indicator for coal rank. The vitrinite maceral is the most common component of coal; due to its high sensitivity to temperature, its mean maximum reflectance determined in polarized light can be used as a geological thermometer and thus as an indicator for the thermal maturity of coal beds (McCartney and Teichmller, 1972).

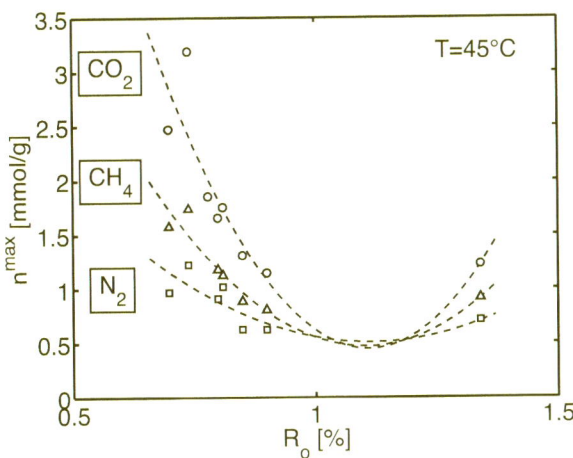

Figure 2.10: Maximum sorption capacity $n^{max}$ of $CO_2$ (○), $CH_4$ (△) and $N_2$ (□) at 45°C as a function of vitrinite reflectance $R_o$. Symbols are experimental points, whereas lines are fitted parabolic curves.

Figure 2.10 shows the maximum sorption capacity $n^{max}$ for $CO_2$, $CH_4$ and $N_2$ as a function of the vitrinite reflectance $R_o$ for eight coal samples. Unfortunately, the data are not homogeneously distributed over the whole range of vitrinite reflectance values, i.e. between 0.7 and 1.3%. However, for all three gases a reduction in sorption capacity with increasing rank can be observed. In a previous study, the surface area of coal, which is known to control the adsorption phenomenon, has been shown to follow a parabolic profile as a function of coal rank (Mahajan, 1991). The experimental points obtained in this study have been fitted with the same function yielding a minimum at about $R_o$=1.1%. The appearance and position of the minimum is in agreement with data reported in other studies (Day et al., 2008a). Further interpretation of this behavior is however precluded in this study by the lack of data over the range of vitrinite reflectance where the minimum appears.

## 2.5 Discussion and concluding remarks

In the light of the results obtained in this work, the issues raised in the introduction regarding the reliability and the use of the sorption data on coal for ECBM applications are discussed in the following.

Comparisons between gas sorption data measured in different laboratories are needed to define a standard procedure for measuring sorption isotherms accurately. This is in fact the only way to be able to give reliable estimates for gas storage in coal seams. Recently, it was reported that $CO_2$ sorption data on Argonne Premium coal samples diverged significantly among laboratories using different measuring techniques both for dried and moisture-equilibrated coal samples (Goodman et al., 2004, 2007). On the contrary, the experimental results presented in this work from two laboratories using two different gravimetric methods showed high reproducibility. This confirms that, when both sample preparation and gas sorption experiments are carried out by carefully following the same procedure, reliable results can indeed be obtained, even if different

## 2.5 Discussion and concluding remarks

measuring techniques are used.

The sorption data of $CO_2$, $CH_4$ and $N_2$ on the different coal samples from several coal mines worldwide presented in this study, show that there is significant difference in terms of sorption capacity among the different coals, which affect their suitability for ECBM purposes. The observed maximum sorption capacity per unit mass of dry coal ranges between 5% and 14% weight for $CO_2$ and between 1% and 3% weight for $CH_4$ and $N_2$. These values provide estimates for the coal bed capacity for $CO_2$ storage and for the maximum theoretical amount of coalbed methane, the so-called maximum Gas In Place ($GIP_{max}$). Being sorption the main mechanism for gas storage in coal seams, high sorption capacity is desirable for both $CO_2$, which can be stored in large amounts, and $CH_4$, which can be recovered. Moreover, the $GIP_{max}$ can be used as a comparison with the actual amount of $CH_4$ in the coal seam: from a storage point of view coal beds with $CH_4$ content close to the $GIP_{max}$ are suitable, since the amount of $CH_4$ which has left the seam is low and therefore the caprock sealing efficiency can be assumed to be high. Finally, the low sorption capacity of $N_2$ makes it suitable for coinjection with $CO_2$ allowing for a faster recovery of $CH_4$ and for a limited reduction of permeability caused by swelling, as shown by modeling studies (Seto et al., 2006; Shi and Durucan, 2006; Bustin et al., 2008) and field tests (Reeves, 2004; Gunter et al., 2004; Yamaguchi et al., 2006).

The relatively high variability in its chemical and physical properties highlights the heterogeneous nature of coal, which results in a different capacity with respect to gas sorption. Correlations between such properties and amount of gas adsorbed are desirable, both for comparing sorption isotherms measured on samples from different coal seams worldwide, and for choosing the most suitable coal seams for ECBM. In agreement with data reported in other studies (Day et al., 2008a), the maximum sorption capacity was found to decrease with increasing vitrinite reflectance and to go through a minimum. Moreover, in the present work it has been observed that for the data on the left hand side of the minimum, i.e. for $R_o < 1$, the difference in sorption capacity between $CO_2$, $CH_4$ and $N_2$ increases with decreasing coal rank. A big difference in sorption capacity is a prerequisite for an efficient displacement of $CH_4$ by the injected $CO_2$. Therefore, from an ECBM point of view, this observation suggests that coal of lower rank should be preferred, in particular if the main goal of the operation is the storage of $CO_2$.

For a quantitative estimation of the storage capacity of the whole coal reservoir the typical structure of coal seams need to be considered, where layers of almost pure coal are interbedded with other rocks. The thickness of both coal and rock layers may vary slightly from place to place, but is of the order of some meters each (Van Krevelen, 1981). Although coal seams are relatively thin, they may extend over large distances, where the different properties usually remain constant (Van Krevelen, 1981). This feature is important in view of the characterization of the coal seam for an ECBM process, in particular in terms of sorption capacity. Direct information about the underground seam structure can in fact only be obtained by analyzing cores obtained from drilling boreholes, a practice which is expensive. It is therefore very important to acquire the maximum amount of data from this sampling. As a typical scenario of data collection, coal cores drilled from a well in Switzerland (Weiach, ZH) and obtained at different depths were investigated. Sorption capacity was found to vary considerably (between 3% and 6% weight) among the samples from different depths. These result highlights that sorption capacity varies considerably with depth and is not homogenously distributed over the reservoir, being this a combination of coal layers, where gas adsorption takes place, but also of transition

## 2.5 Discussion and concluding remarks

layers, where sorption is quite low. This is an important consideration to take into account when estimating the $CO_2$ storage potential of the overall reservoir.

In conclusion, this work has shown that the measurement of supercritical adsorption isotherms represent a scientific and technical challenge. With coal this situation is even more complicated, since, due to the intrinsic variability of the geological processes leading to its formation, it possesses a heterogeneous nature, particularly evident in its chemical and physical properties. Besides, the study of gas sorption on coal is strongly application oriented, since these data are needed to predict the reservoir behavior during an ECBM operation. This requires a quantitative estimation of the $CO_2$ storage capacity of the coal seam, which has to rely on accurate measurement techniques and on a correct interpretation of the obtained data. In this context some interesting issues have been addressed in this study and relevant considerations on the use of the obtained data for ECBM applications have been made. As a step further in the evaluation of the storage potential of coal seams, future research on adsorption should focus on the effect of moisture on the uptake of $CO_2$, since coal seams are naturally saturated with water. Techniques for the measurement of adsorption isotherms on wet samples are in fact not as well established as those on dry samples (see Chapter 8).

## 2.6 Nomenclature

| | |
|---|---|
| $b$ | Langmuir equilibrium constant [cm$^3$/g] |
| $D$ | constant of the Dubinin-Radushkevich isotherm |
| $H$ | Henry's constant (total sorption isotherm) [cm$^3$/g] |
| $k$ | Henry's constant (absorption isotherm) [cm$^3$/g] |
| $m_0^{\text{ads}}$ | Mass of adsorbent (coal) [g] |
| $M_{\text{m}}$ | Molar mass of adsorbate [g/mol] |
| $\mathcal{M}_1$ | Weight at measuring point 1 [g] |
| $\mathcal{M}_1^0$ | Weight at measuring point 1 under vacuum [g] |
| $n^{\text{a}}$ | Molar absolute adsorption [mmol/g] |
| $n^{\text{eas}}$ | Molar excess sorption [mmol/g] |
| $n^{\text{ex}}$ | Molar excess adsorption [mmol/g] |
| $n^{\text{max}}$ | Saturation capacity (Langmuir isotherm) [mmol/g] |
| $n^{\text{s}}$ | Molar absolute absorption [mmol/g] |
| $n^{\text{t}}$ | Total sorption (adsorption + absorption) [mmol/g] |
| $P$ | Pressure [bar] |
| $R$ | Root mean squared error [mmol/g] |
| $\hat{R}$ | Normalized root mean squared error [%] |
| $R_{\text{o}}$ | Vitrinite Reflectance [%] |
| $\rho$ | Density [mmol/cm$^3$] |
| $s^{\text{max}}$ | Saturation capacity (swelling isotherm) [mmol/g] |
| $T$ | Temperature [°C] |
| $V$ | Sorption volume ($V^{\text{a}} + \Delta V^{\text{s}}$) [cm$^3$/g] |
| $\hat{V}$ | Estimated value of $V$ from graphical method [cm$^3$/g] |
| $V^0$ | Volume of lifted metal parts and coal sample [cm$^3$] |
| $V^{\text{a}}$ | Volume of the adsorbed phase [cm$^3$/g] |
| $\Delta V^{\text{s}}$ | Volume of the mixture (coal + imbibed $CO_2$) minus initial sample volume [cm$^3$] |

## 2.6 Nomenclature

**Subscripts and Superscripts**

| | |
|---|---|
| a | adsorption |
| b | bulk |
| s | absorption |
| exp | experiment |
| mod | model |

# Chapter 3

# Sorption on coal: gas mixtures

## 3.1 Introduction

[1]The knowledge of the sorption behavior of a binary mixture of carbon dioxide and methane on coal is the information needed for any study aimed at the description of the displacement phenomenon during ECBM. Moreover, other binary systems, such as carbon dioxide-nitrogen or nitrogen-methane, or even ternary mixtures, may be promising to treat, being the direct injection of a flue gas instead of pure $CO_2$ into a coal seam an option under consideration. In the previous chapter, pure $CO_2$, $CH_4$ and $N_2$ sorption isotherms have been measured on several coal samples from different coal mines worldwide, showing that $CO_2$

---
[1]Part of this work has been published as Ottiger et al. (2008a,b).

adsorbs more than $CH_4$, and $CH_4$ more than $N_2$. This property is of key importance for ECBM application and is confirmed by several multicomponent adsorption studies, which have been more and more undertaken recently (Stevenson et al., 1991; Chaback et al., 1996; Ceglarska-Stefanska and Zarebska, 2005; Shimada et al., 2005; Fitzgerald et al., 2006). Only in two cases, preferential adsorption of $CH_4$ over $CO_2$ on low rank coals has been observed (Busch et al., 2006, 2003). While several studies focused on the binary adsorption of $CO_2$ and $CH_4$, literature about the adsorption of mixtures containing nitrogen is still scarce (Mazumder et al., 2006b; Busch et al., 2007).

A variety of semiempirical isotherm models have been used to describe the experimental adsorption data: for single gas adsorption these include the Langmuir, Toth, Dubinin-Radushkevich and Dubinin-Astakhov isotherms (Clarkson and Bustin, 2000; Bae and Bhatia, 2006; Sakurovs et al., 2007), whereas for multicomponent gas adsorption their combination with the ideal adsorbed solution (IAS) theory (Stevenson et al., 1991; DeGance et al., 1993; Clarkson and Bustin, 2000; Yu et al., 2008b) or the extended Langmuir equation (Arri et al., 1992; Chaback et al., 1996) have been applied. Other methods use an equation of state (EOS), namely a 2-D EOS, such as the Eyring and Virial equation of state (DeGance et al., 1993) or the Zhou-Gasem-Robinson two-dimensional equation of state (Fitzgerald et al., 2005). The relatively simple form of all these equations allow for their direct implementation in reservoir simulators describing ECBM dynamics. An alternative approach uses fluid-fluid and fluid-solid molecular interaction energies and a microscopic description of the pore geometry in the framework of statistical thermodynamics (Sudibandriyo et al., 2003a; Fitzgerald et al., 2006). When quantitative information about the pore size distribution of the coal are incorporated, insights on the behavior of the adsorbed

## 3.1 Introduction

gas in pores of different sizes, particularly under near- and super-critical conditions, can be obtained (Hocker et al., 2003).

In our laboratory, a gravimetric-chromatographic technique has been developed for measuring competitive gas adsorption isotherms (Ottiger et al., 2008a). In particular, binary and ternary sorption experiments with gas mixtures of different composition of $CO_2$, $CH_4$ and $N_2$ have been performed at a temperature of 45°C and at pressures up to 190 bar. All the measured data have been presented and discussed extensively in previous publications (Ottiger et al., 2008a,b). Moreover, the excess sorption isotherms have been successfully described using a lattice density functional theory model based on the Ono-Kondo equations exploiting information about the pore structure of the coal, the adsorbed gases, and the interaction between them. The results clearly showed preferential sorption of $CO_2$ over $CH_4$ and $N_2$, which could be explained by the different interactions energies (fluid-fluid and fluid-solid) between the different components of the gas mixture, thus leading to an enhancement of one adsorbate relative to the others (Kurniawan et al., 2006).

In this chapter, some of the previously published data (Ottiger et al., 2008a,b) are further analyzed in a more application-oriented way, which is therefore more useful for ECBM studies. In particular, the analysis will be carried out by looking at absolute quantities, instead of excess, being this the information needed for gas storage estimations. The results will be then discussed in terms of selectivity, which is one of the most important characteristics for practical applications such as ECBM, where an effective displacement is required. Finally, the predictability of the extended Langmuir equation, the most applied adsorption model in ECBM reservoir simulators, is tested and discussed.

## 3.2 Experimental section

### 3.2.1 Coal characterization

A coal sample from the Monte Sinni coal mine (Carbosulcis, Cagliari, Italy) in the Sulcis Coal Province was used. This sample belongs to the same batch as sample I2 investigated in Chapter 2. The sample was drilled in July 2006 at a depth of about 500 m and preserved in a plastic bottle in air. For the sorption measurements, the coal sample was ground and sieved to obtain particles with diameter between 250 and 355 $\mu$m. Subsequently, it was dried in an oven at 105°C under vacuum for one day. The following pure gases obtained from Pangas (Dagmersellen, Switzerland) were used in this study, namely $CO_2$ and $CH_4$ at purities of 99.995 % and $N_2$ and He at purities of 99.999 %. Binary and ternary gas mixtures of certified compositions were purchased from Pangas (Dagmersellen, Switzerland), that prepared them using $CO_2$, $CH_4$ and $N_2$ at purities of 99.995 %, 99.995 % and 99.9996 %, respectively. The molar compositions of the four carbon dioxide/methane gas mixtures are 20.0, 40.0, 60.0 and 80.0 % $CO_2$, respectively. The molar compositions of the four carbon dioxide/nitrogen mixtures are 10.0, 25.0, 50.0 and 75.0 % $CO_2$, whereas they are 10.0, 25.0, 50.0 and 75.0 % $CH_4$ for the three methane/nitrogen mixtures, respectively. Finally, the ternary mixture carbon dioxide/methane/nitrogen has a molar composition of 33.3 % $CO_2$, 33.3 % $CH_4$ and 33.4 % $N_2$.

### 3.2.2 Experimental technique

All competitive sorption measurements reported in this study were performed in an experimental setup developed and built in-house partially

## 3.2 Experimental section

using commercially available components, a scheme of which is shown in Figure 3.1. A gravimetric-chromatographic method is used for the experiments: the heart of the set-up is a Rubotherm magnetic suspension balance (Bochum, Germany) of the same kind as the one used in Chapter 2 for the pure gas sorption experiments and where a basket containing about 3 g of coal is placed (coal sample 1, $m_{10}^{\text{coal}}$). The balance is connected to an auxiliary cell, where extra adsorbent (about 40 g) can be placed (coal sample 2, $m_{20}^{\text{coal}}$) allowing to amplify the change in the gas-phase composition upon adsorption, thus improving the accuracy of the measuring technique. Between the balance and the auxiliary cell, a circulation pump ensures a homogeneous distribution of the fluid, whose composition can be measured with a gas chromatograph. Details about the experimental setup and the measurement procedure can be found in Ottiger et al. (2008a). However, for the sake of clarity, the most important equations for data reconciliation are reported here.

As in the case of pure gas sorption experiments, the only truly measurable quantity from the balance signal $M_1(\rho^{\text{b}}, T)$ is the so-called excess sorption, which accounts for the effect of both adsorption and absorption, i.e.

$$\begin{aligned} m_1^{\text{eas}}(\rho^{\text{b}}, T) &= m_1^{\text{ex}} + m_1^{\text{s}} - \rho^{\text{b}} \Delta V_1^{\text{s}} \\ &= M_1(\rho^{\text{b}}, T) - M_1^0 + \rho^{\text{b}}(V^{\text{met}} + V_{10}^{\text{coal}}) \end{aligned} \quad (3.1)$$

where $m_1^{\text{ex}}$ is the excess adsorption, and $m_1^{\text{s}} - \rho^{\text{b}} \Delta V_1^{\text{s}}$ is the absorption term corrected for the buoyancy. The right-hand side of Eq.(3.1) contains only measurable variables, i.e. the balance signals $M_1(\rho^{\text{b}}, T)$ and $M_1^0$ measured at the desired conditions and under vacuum, the mass density $\rho^{\text{b}}$ and the sum of the volumes of the suspended metal parts $V^{\text{met}}$ and

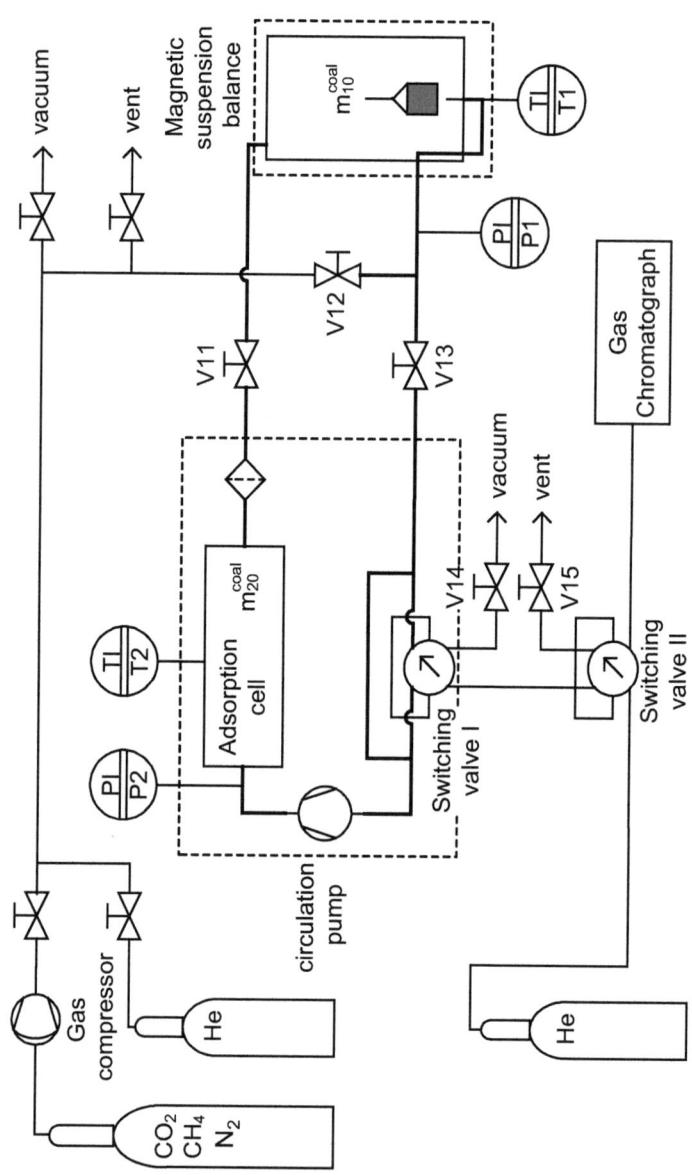

Figure 3.1: Scheme of the setup for competitive adsorption measurements at up to 300 bar and 80°C. For better visibility, the calibrated void volume of the system is connected by thick solid lines.

## 3.2 Experimental section

the initial, unswollen coal sample in the balance $V_{10}^{\text{coal}}$, respectively.
A mass balance over the whole system allows obtaining the total amount of gas fed to the system:

$$\begin{aligned}
m^{\text{feed}}(\rho^{\text{b}}, T) &= m^{\text{a}} + m^{\text{s}} + \rho^{\text{b}}(V_0^{\text{void}} - V^{\text{a}} - \Delta V^{\text{s}}) \\
&= \frac{m_0^{\text{coal}}}{m_{10}^{\text{coal}}} m_1^{\text{eas}} + \rho^{\text{b}} V_0^{\text{void}}
\end{aligned} \quad (3.2)$$

where $m_0^{\text{coal}}$ represents the weight of both coal samples $m_{10}^{\text{coal}} + m_{20}^{\text{coal}}$ and $V_0^{\text{void}}$ the calibrated void volume of the system (with adsorbent). $m^{\text{a}}$ and $m^{\text{s}}$ are respectively the total amount of gas adsorbed and absorbed, and $V^{\text{a}}$ and $\Delta V^{\text{s}}$ the corresponding volumes. It is worth pointing out that $m^{\text{feed}}$ depends on the excess sorption $m_1^{\text{eas}}$, the density $\rho^{\text{b}}$ and the volume $V_0^{\text{void}}$, i.e. quantities that can all be determined experimentally without the need of knowing the exact degree of swelling of the coal. When the mass fractions of the gas mixture fed to the system, $w_i^{\text{feed}}$ and the bulk equilibrium composition, $w_i^{\text{b}}$ are determined through gas chromatography, the individual excess sorption amounts are obtained by making use of mass balances for every component

$$m_i^{\text{eas}} = w_i^{\text{feed}} m^{\text{feed}} - w_i^{\text{b}} \rho^{\text{b}} V_0^{\text{void}}. \quad (3.3)$$

Here, it is worth noting that only the composition $w_i^{\text{feed}}$ and not the exact total amount of gas fed to the system $m^{\text{feed}}$ must be known a priori, since the latter is obtained through the mass balance given by Eq.(3.2).

The experimental results of the gas mixture sorption experiments are then reported in terms of the molar excess sorption $n_i^{\text{eas}}$ of component $i$

per unit mass of coal:

$$n_i^{\text{eas}} = \frac{m_i^{\text{eas}}}{M_{\text{m},i} m_0^{\text{coal}}}, \qquad (3.4)$$

where $M_{\text{m},i}$ represents the molar mass of component $i$ in the mixture. The total molar excess $n^{\text{eas}}$ adsorbed and absorbed per mass of coal is simply obtained by taking the sum over the molar excesses of all $N$ components:

$$n^{\text{eas}} = \sum_{i=1}^{N} n_i^{\text{eas}}. \qquad (3.5)$$

### 3.2.3 Absolute sorption

To be used for gas storage estimations, the experimentally obtained excess isotherms need to be transformed into their corresponding absolute values, which give the actual amount of gas present on the coal. As explained in Chapter 2 for pure gases, the total gas uptake of component $i$, $n_i^{\text{t}}$, can be obtained as follows:

$$n_i^{\text{t}} = n_i^{\text{a}} + n_i^{\text{s}} = n_i^{\text{eas}} + y_i \rho_{\text{m}}^{\text{b}} V \qquad (3.6)$$

where $V$ is defined as $V^{\text{a}} + \Delta V^{\text{s}}$, and $y_i$ is the mole fraction of component $i$ in the bulk phase of molar density $\rho_{\text{m}}^{\text{b}}$. The latter can be obtained as follows

$$y_i = \frac{\frac{w_i}{M_{\text{m},i}}}{\sum_{i=1}^{N} \frac{w_i}{M_{\text{m},i}}} \quad \text{and} \quad \rho_{\text{m}}^{\text{b}} = \frac{\rho^{\text{b}}}{\sum_{i=1}^{N} y_i M_{\text{m},i}} \qquad (3.7)$$

## 3.2 Experimental section

The molar fraction of component $i$ in the adsorbed and absorbed phase can be therefore obtained as $x_i = n_i^t/n^t$, where $n^t$ is the total sorption on the coal. The value of $V$ needed in Eq.(3.6) has been obtained for the pure components $CO_2$, $CH_4$ and $N_2$ by fitting a Langmuir-like equation to the experimentally obtained excess isotherms (see Chapter 2); at a temperature of 45°C, it takes a value of 0.071 ($CO_2$), 0.043 ($CH_4$) and 0.022 cm$^3$/g ($N_2$), respectively. In the case of a gas mixture, the assumption is done here that $V$ changes linearly with composition, i.e. $V = \Sigma_{i=1}^{N} x_i V_i$.

### 3.2.4 Methodology

As shown in Chapter 2, single component gas sorption isotherms can be effectively described by Langmuir-like equations. In this study, the extended Langmuir equation is used to describe the experimentally obtained multi-component gas sorption isotherms on coal:

$$n_i^t = \frac{n_i^{\max} b_i y_i P}{1 + P \sum_{j=1}^{N} b_j y_j} \quad (3.8)$$

where $n_i^t$ is the amount adsorbed and absorbed for component $i$, $y_i$ its molar fraction in the bulk phase and $N$ the number of components; $n_i^{\max}$ and $b_i$ are the saturation capacity per unit mass adsorbent and the Langmuir equilibrium constant, respectively. Their values have been obtained by fitting the single component absolute sorption isotherms presented in Chapter 2 with a Langmuir-like function. Outcome of this procedure is shown in Figure 3.2, where the experimentally obtained $CO_2$, $CH_4$ and $N_2$ sorption isotherms on the Italian coal from the Sulcis Coal Province at 45°C are shown along with the corresponding Langmuir fitted curves as a function of the total pressure $P$.

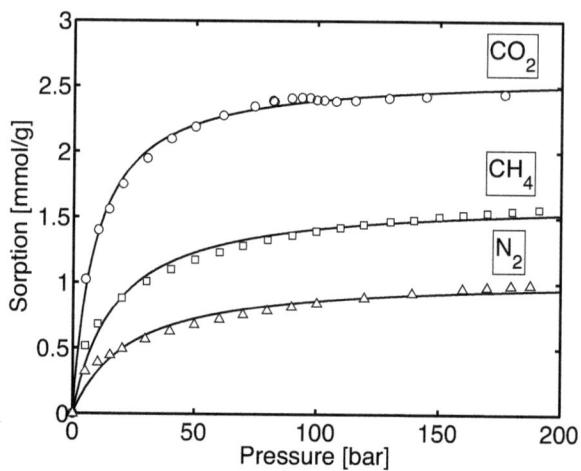

Figure 3.2: Langmuir sorption isotherms at 45°C as a function of pressure for $CO_2$ (○), $CH_4$ (□) and $N_2$ (△) for a coal from the Sulcis coal province. Symbols are experimental points, whereas lines are fitted Langmuir curves.

Values for the fitted parameters are reported in Table 3.1.

Table 3.1: Single component Langmuir equation parameters for $CO_2$, $CH_4$ and $N_2$ sorption on the coal sample used in this study at 45°C.

|  | $n^{max}$ [mmol/g] | $b$ [bar$^{-1}$] |
|---|---|---|
| $CO_2$ | 2.60 | 0.111 |
| $CH_4$ | 1.65 | 0.058 |
| $N_2$ | 1.06 | 0.043 |

In the following sections, only the absolute sorption data are analyzed, as an exhaustive discussion of the gas mixtures excess sorption isotherms

## 3.2 Experimental section

can be found in previous publications (Ottiger et al., 2008a,b). This will allow us to give to the obtained results a more application-oriented imprint, which is therefore very useful for ECBM studies. It is in fact the total gas uptake, i.e. the absolute sorption, the quantity needed for any gas storage estimation in coal. In particular, the sorption data will be presented for a given $(P,T)$ condition as a function of the equilibrium bulk phase composition. This allows for a direct visualization of the actual composition of the adsorbed phase, as the equilibrium bulk composition can be seen as a possible end scenario of a specific ECBM injection scheme (for example pure $CO_2$, pure $N_2$ or mixture of them). This method can be applied to any $(P,T)$ condition; from the data presented in the previous works, only those measured at 45°C are considered and three pressure conditions are chosen, namely 40, 100 and 160 bar; these conditions are representative for a coal seam, such as the one in the Sulcis Coal Province, which lies at about 500 m depth, where temperatures of about 45°C are attained and the hydrostatic pressure is about 50 bar. Moreover, optimal conditions for $CO_2$ storage are expected at larger depths, between 800 and 1600 m, where hydrostatic pressures can be as high as 160 bar. In the case where the pressure at which the experiment was done does not exactly match the needed pressure, the sorption amount is obtained by linear interpolation between the available adjacent measured values.

For practical applications such as ECBM, where an effective displacement is required, the most important characteristic in terms of competitive adsorption is the selectivity of a component with respect to another, i.e.,

$$S_{12} = \frac{x_1/x_2}{y_1/y_2} \tag{3.9}$$

where $x_i$ and $y_i$ are the molar fractions of component $i$ (=1 or 2) in the adsorbed phase and in the bulk phase, respectively. Values greater than unity imply that component 1 is preferentially adsorbed compared to component 2.

## 3.3 Results and discussion

### 3.3.1 Binary mixtures

The binary mixture sorption measurements have been performed at a temperature of 45°C for 11 different gas mixtures of certified composition. The experimental data, obtained through the mass balances described in Section 3.2, are reported in Tables 3.2, 3.3 and 3.4 for the gas mixtures of $CO_2/N_2$, $CO_2/CH_4$ and $CH_4/N_2$, respectively. Together with the sorption values, the tables give also the corresponding binary selectivity $S_{12}$.

It can be seen from these tables that the value of $S_{12}$ is always greater than unity, meaning that component 1 is always preferentially adsorbed compared to component 2. This indicates that $CO_2$ is preferentially adsorbed in the case of $CO_2/N_2$ and $CO_2/CH_4$, whereas $CH_4$ is adsorbed more than $N_2$. Moreover, at a given pressure, selectivity tends to become smaller with increasing bulk molar fraction of the more adsorbed component, in agreement with simulation results of $CO_2/CH_4$ mixtures adsorption in carbons at similar conditions (Kurniawan et al., 2006). The selectivity between $CO_2$ and $CH_4$ is generally lower than the one between $CO_2$ and $N_2$, indicating that in the latter case the competition for sorption is much more in favor of $CO_2$. For the $CH_4/N_2$ gas mixture the selectivity is always between 1 and 2, meaning that $CH_4$ is only slightly preferentially adsorbed compared to $N_2$.

## 3.3 Results and discussion

### CO$_2$ / N$_2$ experiments

Figures 3.3a, b and c, respectively, show the sorption $n_i^t$ of each component $i$ in the mixture per unit mass coal as a function of $y_1$ (CO$_2$) at 45°C and at 40, 100 and 160 bar total gas pressure. The points represent the experimental data, whereas the lines the prediction by the extended Langmuir equation.

Table 3.2: Binary CO$_2$ (1) and N$_2$ (2) sorption data on the coal sample used in this study at 45°C and at 40, 100 and 160 bar.

| P [bar] | $y_1$ | $n_1^t$ [mmol/g] | $n_2^t$ [mmol/g] | $S_{12}$ |
|---|---|---|---|---|
| 40 | 1 | 2.10 | 0 | |
| | 0.72 | 1.68 | 0.12 | 5.56 |
| | 0.44 | 1.37 | 0.25 | 7.13 |
| | 0.19 | 0.85 | 0.49 | 7.54 |
| | 0.06 | 0.42 | 0.59 | 11.17 |
| | 0 | 0 | 0.63 | |
| 100 | 1 | 2.40 | 0 | |
| | 0.74 | 2.06 | 0.15 | 4.85 |
| | 0.47 | 1.73 | 0.26 | 7.51 |
| | 0.22 | 1.18 | 0.51 | 8.29 |
| | 0.07 | 0.69 | 0.60 | 15.19 |
| | 0 | 0 | 0.84 | |
| 160 | 1 | 2.44 | 0 | |
| | 0.74 | 2.13 | 0.22 | 3.40 |
| | 0.48 | 1.84 | 0.32 | 6.24 |
| | 0.23 | 1.31 | 0.57 | 7.75 |
| | 0.08 | 0.76 | 0.71 | 12.38 |
| | 0 | 0 | 0.96 | |

The first, obvious, observation is that the sorption of each component $i$ increases with increasing concentration of the specific component in the bulk phase. Moreover, it can be seen that the model is able to describe the experimental binary data fairly well at each pressure and over the

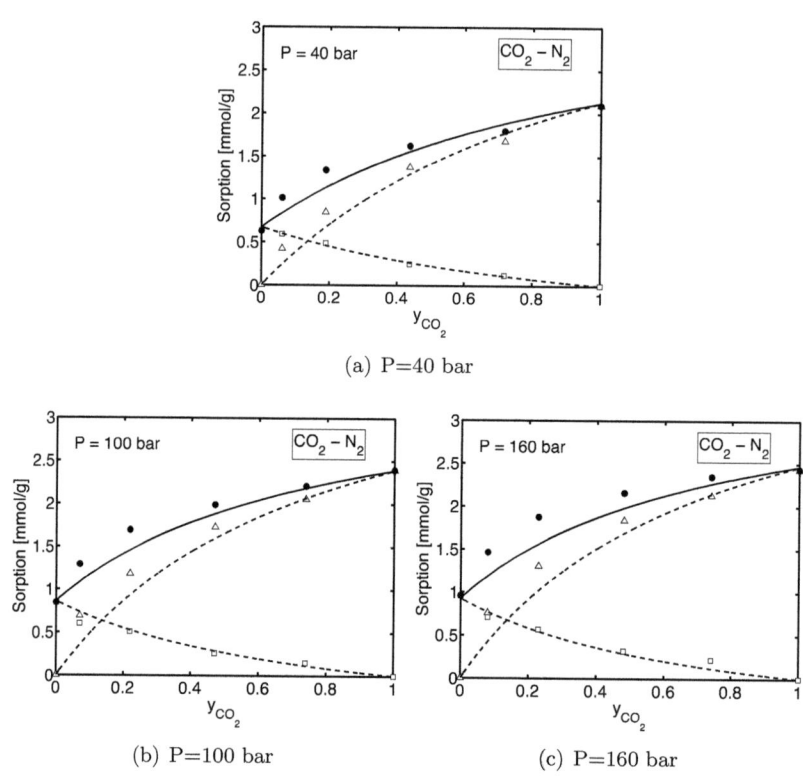

Figure 3.3: Binary high-pressure sorption isotherms of $CO_2$ (1) and $N_2$ (2) on the Sulcis coal at 45°C at 40 (a), 100 (b) and 160 bar (c) as a function of gas composition $y_1$ ($CO_2$). Symbols are experimental points, whereas lines are the predicted extended Langmuir curves. Symbols: $CO_2$ (△), $N_2$ (□), total (●).

## 3.3 Results and discussion

whole composition range. Only at low $y_1$ the $CO_2$ sorption tends to be underpredicted. Considering that the model is fully predictive once it has been fitted to the pure sorption data, it provides a good description of the experimental results. Interestingly, the gas mixture $CO_2/N_2$ should contain at least 75 % of $N_2$ to have significant adsorption of $N_2$, which represents less than 10 % of the total adsorption at $N_2$ concentrations below 50 %. This shows that $CO_2$ is strongly preferentially adsorbed with respect to $N_2$, which is an interesting property in view of a possible direct use of flue gas instead of pure $CO_2$ in a ECBM operation.

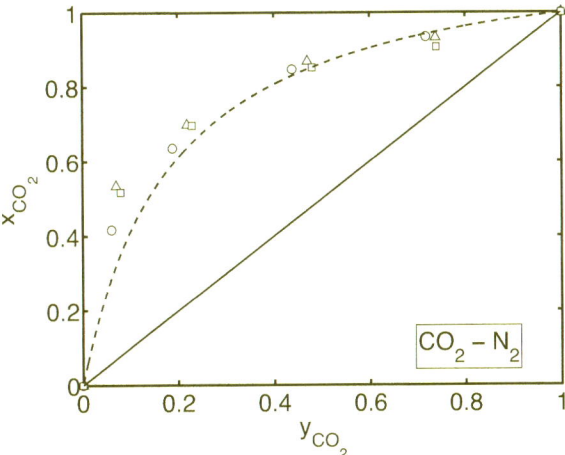

Figure 3.4: x-y diagram for $CO_2/N_2$ mixture at 45°C at three different pressures 40 (○), 100 (△) and 160 bar (□). Symbols are experimental points, whereas lines are extended Langmuir predicted curves.

This behavior can be easily visualized in the x-y diagram of Figure 3.4, which shows that over the whole composition range the mole fraction of $CO_2$ in the adsorbed phase is greater than in the bulk phase. The data are plotted for three different pressures, but no clear trend with pressure can be observed. Again it can be seen that the extended Langmuir model is able to reproduce satisfactorily the experimentally obtained sorption data.

Table 3.3: Binary $CO_2$ (1) and $CH_4$ (2) sorption data on the coal sample used in this study at 45°C and at 40, 100 and 160 bar.

| $P$ [bar] | $y_1$ | $n_1^t$ [mmol/g] | $n_2^t$ [mmol/g] | $S_{12}$ |
|---|---|---|---|---|
| 40 | 1 | 2.10 | 0 | |
| | 0.77 | 1.67 | 0.10 | 4.68 |
| | 0.56 | 1.40 | 0.28 | 3.96 |
| | 0.35 | 1.11 | 0.46 | 4.43 |
| | 0.16 | 0.67 | 0.70 | 5.03 |
| | 0 | 0 | 1.10 | |
| 100 | 1 | 2.40 | 0 | |
| | 0.79 | 2.01 | 0.17 | 3.11 |
| | 0.58 | 1.74 | 0.28 | 4.48 |
| | 0.38 | 1.38 | 0.56 | 4.02 |
| | 0.18 | 0.89 | 0.86 | 4.71 |
| | 0 | 0 | 1.40 | |
| 160 | 1 | 2.44 | 0 | |
| | 0.8 | 2.10 | 0.21 | 2.50 |
| | 0.59 | 1.80 | 0.37 | 3.39 |
| | 0.39 | 1.44 | 0.66 | 3.42 |
| | 0.19 | 0.96 | 0.93 | 4.41 |
| | 0 | 0 | 1.52 | |

## $CO_2$ / $CH_4$ experiments

The molar absolute sorption isotherms of each component $i$ in the mixture per unit mass of coal as a function of the fluid composition are shown in Figure 3.5 at the three pressures considered (40, 100 and 160 bar). The experimental data are well represented by the extended Langmuir model over the whole composition range and at each pressure. As expected,

## 3.3 Results and discussion

the effect of preferential adsorption of carbon dioxide over methane is less strong than for $CO_2/N_2$ due to the stronger sorption of $CH_4$ with respect to $N_2$. In Figure 3.6 (x-y diagram), in fact, the experimental data together with the curve predicted by the extended Langmuir model show a less pronounced curvature compared to the $CO_2/N_2$ case.

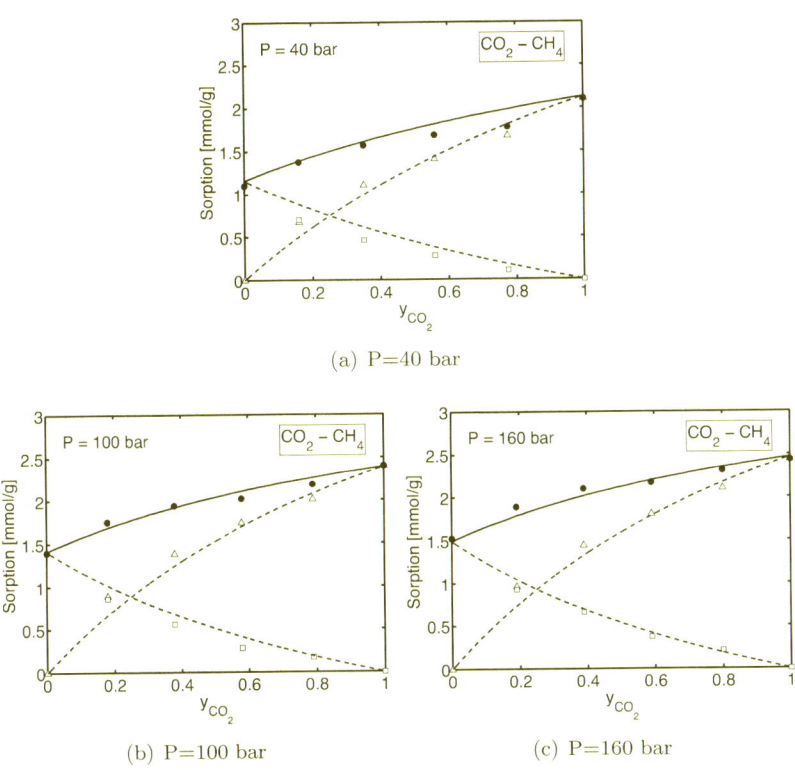

(a) P=40 bar

(b) P=100 bar

(c) P=160 bar

Figure 3.5: Binary sorption isotherms of $CO_2$ (1) and $CH_4$ (2) on the Sulcis coal at 45°C at 40 (a), 100 (b) and 160 bar (c) as a function of gas composition $y_1$ ($CO_2$). Symbols are experimental points, whereas lines are the predicted extended Langmuir curves. Symbols: $CO_2$ (△), $CH_4$ (□), total (●).

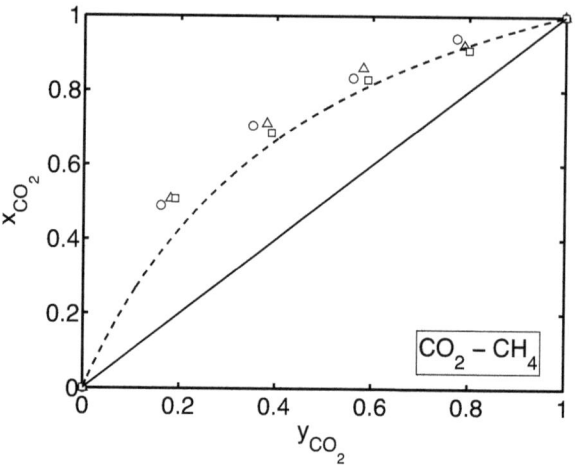

Figure 3.6: x-y diagram for $CO_2/CH_4$ mixture at 45°C at three different pressures 40 (○), 100 (△) and 160 bar (□). Symbols are experimental points, whereas lines are extended Langmuir predicted curves.

#### $CH_4$ / $N_2$ experiments

In the case of binary methane-nitrogen mixtures, Figure 3.7 shows the molar sorption $n_i^t$ per unit mass of coal as a function of $y_1$ ($CH_4$) at 45°C and 40, 100 and 160 bar total gas pressure. It can be observed that the model is able to describe the experimental binary data very well; at an equimolar feed composition of 50 % $CH_4$ and 50 % $N_2$, the sorption of $N_2$ represents at least 30 % of the total sorption at the three pressures studies. Therefore, $CH_4$ is only slightly preferentially adsorbed over $N_2$. As a consequence, on the x-y diagram (Figure 3.8), the experimental points and the model predicted curve are much closer to the line representing no selectivity between one or the other component (solid line).

## 3.3 Results and discussion

Table 3.4: Binary CH$_4$ (1) and N$_2$ (2) sorption data on the coal sample used in this study at 45°C and at 40, 100 and 160 bar.

| $P$ [bar] | $y_1$ | $n_1^t$ [mmol/g] | $n_2^t$ [mmol/g] | $S_{12}$ |
|---|---|---|---|---|
| 40 | 1 | 1.10 | 0 | |
| | 0.74 | 0.83 | 0.18 | 1.64 |
| | 0.48 | 0.58 | 0.32 | 1.99 |
| | 0.24 | 0.29 | 0.45 | 2.07 |
| | 0 | 0 | 0.63 | |
| 100 | 1 | 1.40 | 0 | |
| | 0.75 | 1.06 | 0.22 | 1.61 |
| | 0.49 | 0.74 | 0.41 | 1.88 |
| | 0.24 | 0.39 | 0.59 | 2.10 |
| | 0 | 0 | 0.84 | |
| 160 | 1 | 1.52 | 0 | |
| | 0.75 | 1.17 | 0.24 | 1.62 |
| | 0.49 | 0.83 | 0.44 | 1.96 |
| | 0.25 | 0.44 | 0.66 | 2.00 |
| | 0 | 0 | 0.96 | |

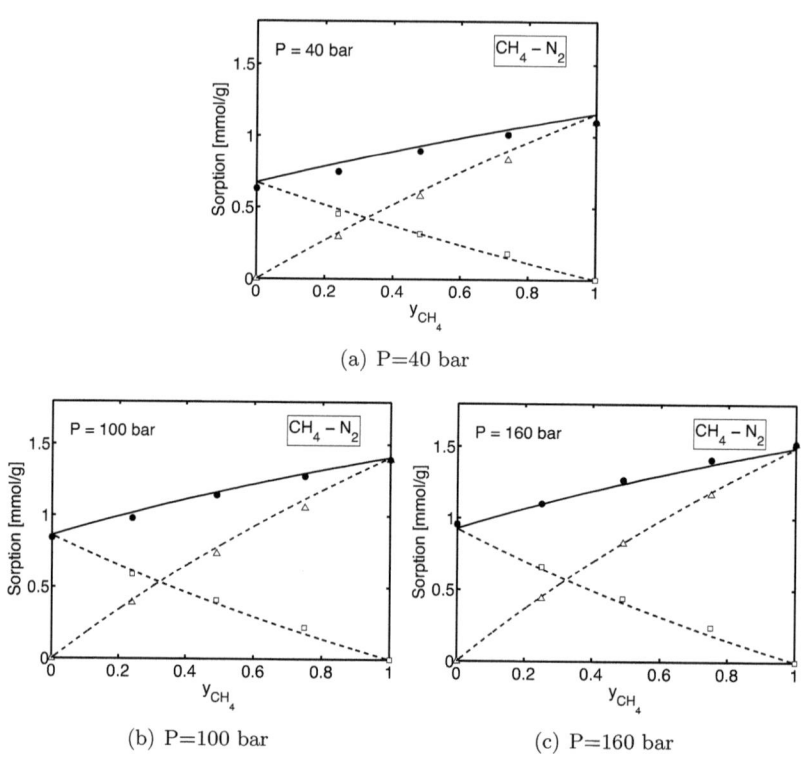

Figure 3.7: Binary high-pressure sorption isotherms of $CH_4$ (1) and $N_2$ (2) on the Sulcis coal at 45°C at 40 (a), 100 (b) and 160 bar (c) as a function of gas composition $y_1$ ($CH_4$). Symbols are experimental points, whereas lines are the predicted extended Langmuir curves. Symbols: $CH_4$ (△), $N_2$ (□), total (●).

## 3.3 Results and discussion

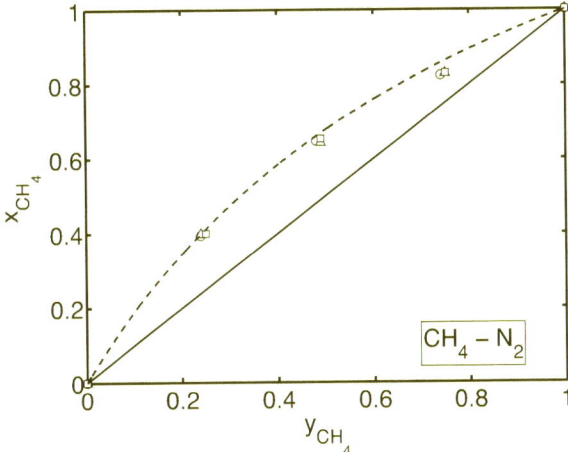

Figure 3.8: x-y diagram for $CH_4/N_2$ mixture at 45°C at three different pressures 40 (○), 100 (△) and 160 bar (□). Symbols are experimental points, whereas lines are extended Langmuir predicted curves.

## 3.3.2 Ternary mixture

The ternary adsorption measurements have been performed at a temperature of 45°C and pressures up to 180 bar for a gas mixture of certified feed composition of 33.3 % $CO_2$, 33.3 % $CH_4$ and 33.4 % $N_2$. The experimental data are reported in Table 3.5.

Table 3.5: Ternary $CH_4$ (1) and $N_2$ (2) and $N_2$ (3) sorption data on the coal sample used in this study at 45°C. The gas phase composition was determined by gas chromatography.

| $P$ [bar] | $y_1$ | $y_2$ | $n_1^t$ [mmol/g] | $n_2^t$ [mmol/g] | $n_3^t$ [mmol/g] |
|---|---|---|---|---|---|
| 10.3 | 0.21 | 0.38 | 0.57 | 0.18 | 0.12 |
| 20.2 | 0.24 | 0.37 | 0.79 | 0.22 | 0.15 |
| 40.6 | 0.27 | 0.36 | 1.01 | 0.26 | 0.18 |
| 60.8 | 0.29 | 0.35 | 1.15 | 0.27 | 0.19 |
| 83.1 | 0.30 | 0.35 | 1.24 | 0.29 | 0.21 |
| 103.0 | 0.30 | 0.35 | 1.30 | 0.29 | 0.20 |
| 122.8 | 0.31 | 0.34 | 1.34 | 0.30 | 0.22 |
| 143.0 | 0.31 | 0.34 | 1.37 | 0.30 | 0.22 |
| 160.9 | 0.31 | 0.34 | 1.37 | 0.32 | 0.24 |
| 180.4 | 0.32 | 0.34 | 1.38 | 0.32 | 0.25 |

Figure 3.9 shows the experimentally obtained sorption isotherms $n_i^t$ per unit mass of coal as a function of pressure $P$ and compares it with the prediction by the extended Langmuir equation. It can observed that also in this case the preferential adsorption reflects the behavior observed in the previous sections for the binary sorption measurements, i.e. in the order $CO_2 > CH_4 > N_2$. The prediction form the Langmuir equation are less satisfactory than for the binary cases, being $CO_2$ systematically underestimated and $CH_4$ overestimated over the whole range of pressure. However, considering that the Langmuir equation is fully predictive, since it is based only on the pure component Langmuir constants,

## 3.4 Concluding remarks

we believe that it still represents a valuable option for the description of multi-component sorption isotherms on coal.

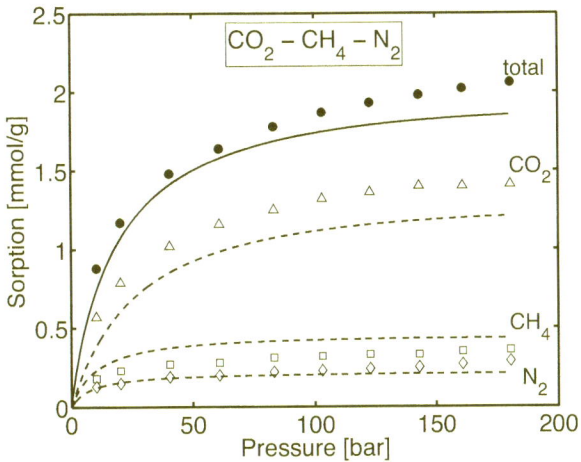

Figure 3.9: Sorption amount $n_i^t$ of component $i$ per unit mass of coal at 45°C as a function of the total pressure, $P$. Feed composition of ternary mixture: 33.3% $CO_2$, 33.3% $CH_4$, 33.4% $N_2$. Symbols: experimental data; Lines: extended Langmuir equation.

## 3.4 Concluding remarks

A comprehensive set of experimental data of pure, binary and ternary sorption of $CO_2$, $CH_4$ and $N_2$ on a dried coal sample from the Sulcis Coal Province (Sardinia, Italy) has been measured at a temperature of 45°C and pressures up to 180 bar using a gravimetric-chromatographic technique and presented in previous studies (Ottiger et al., 2008a,b). In this work, an alternative approach has been proposed to analyze some of the published data, allowing us to present them in a way which is more useful for ECBM studies. In particular, absolute sorption isotherms have been obtained and discussed in terms of selectivity. Moreover, the extended Langmuir equation has been used to describe the data. Considering that the model is fully predictive once it has been fitted to the pure sorption data, it provides a satisfactory description of the experimental results. This approach is very effective when a model has to be developed that can be directly implemented in a process simulator. However, if the goal is to achieve a profound understanding of the thermodynamics of competitive adsorption, other more evolved models have to be used, has it was shown in a previous study, where a lattice density functional theory model based on the Ono-Kondo equations exploiting information about the pore structure of the coal, the adsorbed gases, and the interaction between them has been applied to the same data presented here (Ottiger et al., 2008b).

## 3.5 Nomenclature

| | |
|---|---|
| $b$ | Langmuir equilibrium constant [cm$^3$/g] |
| $m^{\mathrm{a}}$ | Mass adsorbed [g] |
| $m^{\mathrm{eas}}$ | Excess mass adsorbed and sorbed [g] |
| $m^{\mathrm{ex}}$ | Excess mass adsorbed [g] |
| $m^{\mathrm{feed}}$ | Amount of gas fed to the system [g] |
| $m^{\mathrm{s}}$ | Mass absorbed [g] |
| $M_1$ | Weight at measuring point 1 [g] |
| $M_1^0$ | Weight at measuring point 1 under vacuum [g] |
| $M_{\mathrm{m}}$ | Molar mass of adsorbate [g/mol] |
| $n^{\mathrm{a}}$ | Moles adsorbed per unit mass adsorbent [mmol/g] |
| $n^{\mathrm{eas}}$ | Molar excess sorption per unit mass adsorbent [mmol/g] |
| $n^{\mathrm{max}}$ | saturation capacity per unit mass coal [mmol/g] |
| $n^{\mathrm{s}}$ | Moles absorbed per unit mass adsorbent [mmol/g] |
| $n^{\mathrm{t}}$ | Total gas uptake per unit mass adsorbent [mmol/g] |
| $P$ | Pressure [bar] |
| $\rho^{\mathrm{b}}$ | Bulk density [g/cm$^3$] |
| $S_{12}$ | Selectivity between components 1 and 2 [-] |
| $\Delta V^{\mathrm{s}}$ | Volume of sorbed phase [cm$^3$] |
| $V$ | Volume of the adsorbed and absorbed phase [cm$^3$/g] |
| $V^{\mathrm{met}}$ | Volume of lifted metal parts and coal sample [cm$^3$] |
| $V_0^{\mathrm{coal}}$ | Initial (unswollen) coal sample volume [cm$^3$] |
| $V_0^{\mathrm{void}}$ | Initial system void volume [cm$^3$] |
| $w$ | Weight fraction in the bulk phase [-] |
| $x$ | Molar fraction in the sorbed phase [-] |
| $y$ | Molar fraction in the bulk phase [-] |

# Chapter 4

# Swelling of coal

## 4.1 Introduction

Uptake and release of many gases and liquids are associated with swelling and shrinking of coal, respectively (Larsen, 2004). The phenomenon of coal swelling has been introduced after the first attempts aimed at the determination of the coal surface area. $CO_2$ has been proposed as a measuring gas since, through dissolution into the coal matrix, it is able to reach both open and closed pores of coal, allowing to obtain meaningful estimates of the surface area (Mahajan, 1991). Those experiments are conducted at low pressures and only recently the interest moved towards higher pressures, i.e. at pressures relevant for $CO_2$ storage. The dilatometric, optical or strain measurement methods are the most common techniques used to measure the volume changes of coal samples of given shape upon exposure to a gas at high pressure for several days (Reucroft and Sethuraman, 1987; Harpalani and Chen, 1995; Ottiger et al., 2008b).

A summary of the studies reported in the literature together with the corresponding applied experimental conditions is given in Table 4.1.

According to the dual nature of $CO_2$ uptake, i.e. surface adsorption and absorption into the solid matrix, swelling can be interpreted in two complementary ways. On the one hand, adsorption induces a change of the coal specific surface energy, which can be compensated by the elastic energy change associated to the volume change (Scherer, 1986; Pan and Connell, 2007). On the other hand, as a glassy, strained, cross-linked macromolecular system, coal undergoes structural changes in the presence of high pressure $CO_2$ that can be explained only by penetration of $CO_2$ into the coal matrix (Larsen, 2004; Karacan, 2003). Through this mechanism, $CO_2$ uptake may lead to weakening and plasticization phenomena, as well as to changes of coal mechanical properties, e.g. its softening temperature and its Young's elastic modulus, possibly over the long time horizon of $CO_2$ storage (Larsen, 2004; Van Krevelen, 1981; Viete and Ranjith, 2006; Wang et al., 2007).

In this chapter, two coal samples obtained from two different coal mines in Italy have been investigated in terms of their volumetric behavior. In particular, the effect of exposing a coal disc to an atmosphere of an adsorbing gas such as $CO_2$, $CH_4$ and $N_2$ as well as to the inert helium is investigated at temperatures of 45°C and pressures up to 140 bar and the resulting swelling is determined using a high-pressure view cell. Finally, a Langmuir-like equation is proposed to describe the obtained isotherms.

## 4.1 Introduction

Table 4.1: Studies reporting swelling measurements on coal. Maximum pressure, $P_{max}$ is given in bar.

| Coal origin | Method | Gas | $P_{max}$ | Ref. |
|---|---|---|---|---|
| Australia | optical | $CO_2$ | 150 | Day et al. (2008b) |
| Canada | strain | $CO_2/CH_4/N_2$ | 50 | Cui et al. (2007) |
| France | strain | $CO_2/CH_4/He$ | 55 | Durucan et al. (2008) |
| Germany | strain | $CO_2/CH_4/He$ | 55 | Durucan et al. (2008) |
| Sulcis Coal Province (Italy) | optical | $CO_2/CH_4/N_2/He$ | 140 | Ottiger et al. (2008b) |
| South Island (New Zealand) | strain | $CO_2/CH_4/N_2/He$ | 40 | St. George and Barakat (2001) |
| Poland | dilatometric | $CO_2$ | 40 | Ceglarska-Stefanska and Czaplinski (1993) |
| United Kingdom | strain | $CO_2/CH_4/He$ | 55 | Durucan et al. (2008) |
| San Juan Basin, USA | strain | $CH_4/He$ | 104 | Harpalani and Chen (1995) |
| Kentucky, USA | dilatometric | $CO_2$ | 15 | Reucroft and Sethuraman (1987) |
| USA | dilatometric | $CO_2/N_2/He$ | 48 | Walker et al. (1988) |

## 4.2 Experimental section

### 4.2.1 Materials and experimental setup

Two coal samples from Italy were used in this study. Sample I1 was taken from the Ribolla Coal Mine (Grosseto, Italy), whereas sample I2 comes from the Monte Sinni coal mine (Carbosulcis, Cagliari, Italy) in the Sulcis Coal Province. Main properties of these samples are reported in Table 4.2. From the coal blocks, a coal disc of about 22 mm in diameter was drilled with its two faces cut parallel. A disc shape was chosen as a compromise between measurement precision and equilibration time when the coal is exposed to a fluid. Prior to the swelling experiments, the coal discs were dried in an oven at a temperature of 105°C for two days in order to remove any pre-adsorbed moisture.

Table 4.2: Properties of the two Italian coal samples investigated.

| Sample | I1 | I2 |
|---|---|---|
| Moisture (%) | 7.80 | 5.32 |
| Volatile Matter (%) | 30.99 | 40.25 |
| Fixed Carbon (%) | 50.09 | 45.72 |
| Ash (%) | 11.12 | 8.71 |
| $R_o$ (%) | 0.74 | 0.70 |
| Density (g/cm$^3$) | 1.435 | 1.370 |

The swelling experiments were performed in a high-pressure view cell, which has been used in the past to study the expansion of polymers (Rajendran et al., 2005; Bonavoglia et al., 2006; Pini et al., 2008). The view cell is a cylindrical vessel and has a volume of about 50 cm$^3$. Circular sapphire windows, which are orthogonal to the axis of the cylinder, are mounted at its two ends. The view cell is immersed in a thermostated water bath and is equipped with a pressure transducer with a resolution

## 4.2 Experimental section

of 1 bar. The measurements are done by direct visualization, i.e. digital stop motion images of the swollen coal sample were acquired by a digital photocamera (see Figure 4.1).

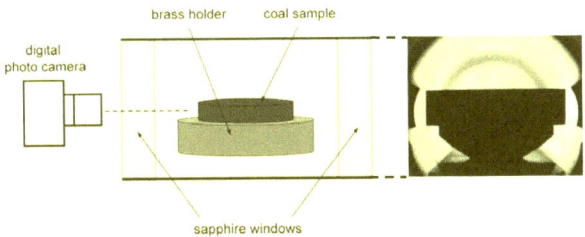

Figure 4.1: Schematic of the high-pressure view cell used for the swelling experiments (left) and obtained image of the coal sample (right).

### 4.2.2 Experimental procedure

The experimental procedure has been described in detail in previous publications (Rajendran et al., 2005; Pini et al., 2008). Nevertheless, the most important steps of the data reconciliation as well as the equations to determine the degree of swelling are briefly summarized in the following. In the swelling experiments the coal disc is placed on a brass holder in the high pressure view cell, as depicted in Figure 4.1. Beside keeping the coal sample in horizontal position, the brass holder is the reference for evaluating the diameter of the swollen coal sample from the digital picture. Such quantity is estimated by comparing its size to that of the brass holder, being this last quantity not affected by fluid pressure. The cell is brought to the desired temperature, is flushed with the fluid to be used and then filled to the required pressure. The coal disc is allowed to expand for two days to reach equilibrium conditions before a picture is taken and the diameter of the disc is determined using a commercial image analysis software. The swelling $s$ is defined as follows

$$s = \frac{V^{\text{coal}} - V_0^{\text{coal}}}{V_0^{\text{coal}}} = \frac{\Delta V^s}{V_0^{\text{coal}}} \qquad (4.1)$$

where $V_0^{\text{coal}}$ and $V^{\text{coal}}$ are initial and final volumes of the coal disc, respectively. The difference $V^{\text{coal}} - V_0^{\text{coal}}$ corresponds to the volume of the sorbed phase $\Delta V^s$, i.e. the volume increase of the coal due to sorption. The volume change of the disc upon swelling can be calculated as follows:

$$s(\rho^b, T) = \frac{\frac{\pi}{4}d^2(\rho^b, T)h(\rho^b, T)}{\frac{\pi}{4}d_0^2 h_0} - 1 \qquad (4.2)$$

where $d$ and $h$ are respectively the diameter and the thickness of the coal disc at the given density $\rho^b$ and temperature $T$, and the subscript 0 refers to the initial unswollen state. If isotropic expansion is assumed, i.e.

$$\frac{d(\rho^b, T)}{d_0} = \frac{h(\rho^b, T)}{h_0} \qquad (4.3)$$

the final expression for the volumetric expansion is obtained,

$$s(\rho^b, T) = \frac{d^3(\rho^b, T)}{d_0^3} - 1 \qquad (4.4)$$

Finally, the pressure inside the view cell is increased to a higher level and the described procedure is repeated, allowing to obtain a complete swelling isotherm.

## 4.2 Experimental section

As anticipated in Chapter 2, the experimental swelling data can be described as a function of pressure by Langmuir-like equations, i.e.

$$s(\rho^{\text{b}}, T) = \frac{s^{\max} b_{\text{s}} P}{1 + b_{\text{s}} P} \tag{4.5}$$

where $s^{\max}$ and $b_{\text{s}}$ are temperature dependent, gas and coal specific constants (Cui et al., 2007; Levine, 1996).

### 4.2.3 Results and discussion

Using the view cell, measurements of swelling were carried out using the three adsorbing fluids ($CO_2$, $CH_4$ and $N_2$) and the inert helium at a temperature of 45°C and up to a pressure of 130 bar. All the experimental data of swelling are reported in Table 4.3 and 4.4 for coal sample I1 and I2, respectively. These data have been obtained by letting the sample equilibrate for about two days at the given pressure level. Figure 4.2 shows the isotropic swelling $s$ as a function of time at different pressures for the case of coal sample I2; it can be seen that two days are indeed enough to ensure that sorption and swelling equilibrium have been reached.

In Figure 4.3 and 4.4 the isotropic swelling $s$ is shown as a function of the pressure $P$ for $CO_2$, $CH_4$ and $N_2$ at a temperature of 45°C for sample I1 and I2, respectively. Moreover, for coal sample I2, the obtained swelling has been compared to the corresponding swelling under He. The direct visualization method applied in this work allowed us calculating the swelling with a precision shown by the error bars in Figure 4.3 and 4.4, accounting for the fact that the diameter of the coal disc can only be determined with a precision of 1 pixel.

Together with the experimental points are shown the Langmuir fitted

Table 4.3: Experimental swelling data of pure $CO_2$, $CH_4$, $N_2$ on Ribolla coal (I1).

| Fluid | $P$ [bar] | $s$ [-] |
|---|---|---|
| $CO_2$ | 30 | 0.0052 |
|  | 58 | 0.0113 |
|  | 89 | 0.0158 |
|  | 117 | 0.0183 |
| $CH_4$ | 30 | 0.0033 |
|  | 60 | 0.0063 |
|  | 91 | 0.0078 |
| $N_2$ | 30 | 0.0010 |
|  | 41 | 0.0005 |
|  | 63 | 0.0006 |
|  | 93 | 0.0025 |
|  | 124 | 0.0023 |

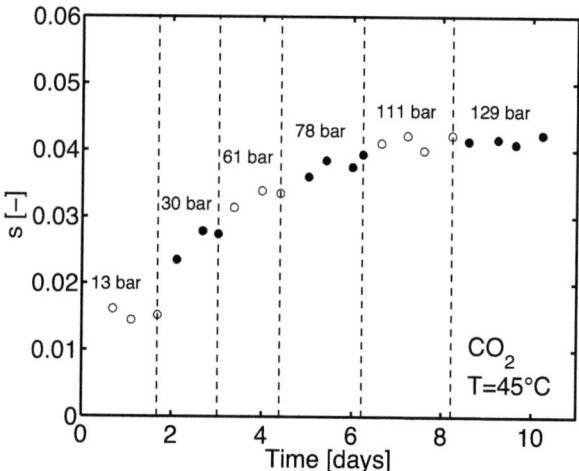

Figure 4.2: Swelling of an unconstrained dry disc (Sulcis coal) as a function of the time when exposed to $CO_2$ at 45°C. Swelling is assumed to be isotropic.

## 4.2 Experimental section

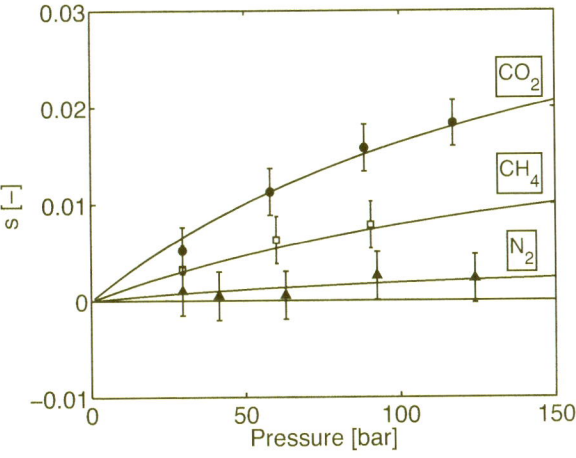

Figure 4.3: Swelling of an unconstrained dry disc (Ribolla coal), as a function of the pressure, $P$, of $CO_2$, $CH_4$ and $N_2$ at 45°C. Swelling is assumed to be isotropic.

Table 4.4: Experimental swelling data of pure $CO_2$, $CH_4$, $N_2$ and He on a coal disc from the Sulcis Coal Province (I2).

| Fluid | $P$ [bar] | $s$ [-] | Fluid | $P$ [bar] | $s$ [-] | Fluid | $P$ [bar] | $s$ [-] |
|---|---|---|---|---|---|---|---|---|
| $CO_2$ | 13 | 0.015 | | 35 | 0.011 | | 53 | 0.006 |
| | 27 | 0.025 | | 41 | 0.016 | | 70 | 0.006 |
| | 30 | 0.027 | | 56 | 0.014 | | 81 | 0.006 |
| | 44 | 0.030 | | 76 | 0.017 | | 90 | 0.005 |
| | 60 | 0.034 | | 78 | 0.020 | | 113 | 0.006 |
| | 61 | 0.034 | | 97 | 0.017 | | 115 | 0.005 |
| | 78 | 0.039 | | 114 | 0.022 | | 129 | 0.006 |
| | 94 | 0.038 | | 116 | 0.020 | | 129 | 0.008 |
| | 99 | 0.037 | $N_2$ | 10 | -0.001 | He | 14 | 0.002 |
| | 111 | 0.042 | | 20 | 0.001 | | 40 | 0.000 |
| | 112 | 0.038 | | 22 | 0.001 | | 69 | -0.001 |
| | 128 | 0.042 | | 31 | 0.002 | | 99 | -0.001 |
| $CH_4$ | 17 | 0.009 | | 45 | 0.004 | | 121 | -0.003 |

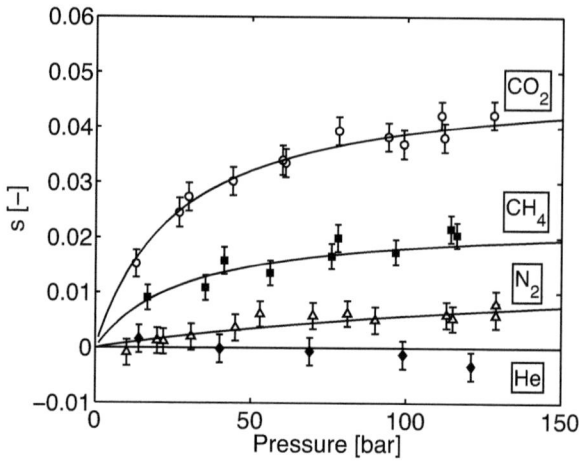

Figure 4.4: Swelling of an unconstrained dry disc (Sulcis Coal) as a function of the pressure, $P$, of $CO_2$, $CH_4$, $N_2$ and He at 45°C. Swelling is assumed to be isotropic.

curves, whose parameters are summarized in Table 4.5. It can be seen that in agreement with several other studies, also in this case the swelling can be effectively described by Langmuir-like curves. In both cases the extent of swelling increases monotonically with pressure up to a few percents for adsorbing gases, with $CO_2$ swelling coal more than $CH_4$ that swells more than $N_2$, whereas for helium, a non-adsorbing gas, volume changes are negligible. Similar observations are reported in the literature for other coal samples (Day et al., 2008b; Cui et al., 2007; St. George and Barakat, 2001). In view of an ECBM operation, these outcomes indicate that the displacement of $CH_4$ by $CO_2$ would lead to a net coal swelling, whereas methane displacement by $N_2$ to a net shrinking. To which extent this volumetric behavior influence the gas flow dynamics will be discussed further on in this thesis.

It can be also seen that for all the fluids used, coal sample I2 shows a

## 4.2 Experimental section

larger swelling degree compared to sample I1. This behavior is somehow in contrast with the outcomes of the gas sorption experiments, where coal sample I1 showed a greater capacity for $CO_2$ compared to sample I2. In the case of $CH_4$ and $N_2$ the adsorbed amounts were similar. Also in previous studies, it has been concluded that coals with high sorption capacity are not necessarily high swelling coals (Day et al., 2008b). This observation suggests that the interaction between the mechanisms of swelling and sorption is not trivial, but that other phenomena, such as strain response to stress, may come into play.

It is worth pointing out, that repeated exposure to the swelling gas doesn't affect irreversibly the coal, since the sample was returning to its original size when the pressure was released, in agreement with previous studies (Cui et al., 2007; Day et al., 2008b). However, recently it was reported that repeated $CO_2$ swelling measurements on coal showed greater changes (30 to 70%) in the direction perpendicular to the bedding plane than in that parallel to it (Day et al., 2008b). In another study, the observed differences between the two directions were very limited (Levine, 1996). The results from the first case suggest that the anisotropic nature of the coal remains unchanged upon repeated exposure to the high-pressure gas, which is different from the behavior observed when organic solvents are used, where after the first exposure the coal behaved isotropically (Larsen et al., 1997; Larsen, 2004). It is clear that further measurements are needed to clarify whether one or the other conclusion can be drawn. However, we believe that the data presented in Figure 4.3 and 4.4 are useful and the assumption of isotropic expansion acceptable, being the error associated to the experimental technique similar to the difference due to assuming in the two mentioned works either isotropic or anisotropic behavior.

It is very common for polymer systems to represent the measured

Table 4.5: Langmuir model parameters for $CO_2$, $CH_4$ and $N_2$ adsorption and swelling on coal samples I1 and I2 at 45°C.

| Sample | Ribolla (I1) | | | Sulcis (I2) | | |
| --- | --- | --- | --- | --- | --- | --- |
| Fluid | $CO_2$ | $CH_4$ | $N_2$ | $CO_2$ | $CH_4$ | $N_2$ |
| $b_s$ [Pa-1] | $5.91 \times 10^{-8}$ | $4.88 \times 10^{-8}$ | $6.02 \times 10^{-8}$ | $3.8 \times 10^{-7}$ | $3.47 \times 10^{-7}$ | $5.19 \times 10^{-8}$ |
| $s^{max}$ [-] | 0.044 | 0.024 | 0.005 | 0.049 | 0.023 | 0.017 |
| $b_p$ [Pa-1] | $5.18 \times 10^{-7}$ | $3.87 \times 10^{-7}$ | $1.18 \times 10^{-7}$ | $1.25 \times 10^{-6}$ | $6.27 \times 10^{-7}$ | $1.4 \times 10^{-7}$ |
| $n^{max}$ [mmol/g] | 3.43 | 1.69 | 1.38 | 2.49 | 1.56 | 1.52 |

swelling data as a function of the concentration of the sorbed fluid (Bonavoglia et al., 2006; Pini et al., 2007). Being dependent on the physical properties of the polymer and on the polymer-solvent interactions, the obtained curve is system specific, but qualitatively it can be described with a S-shape function which can be divided into three characteristic regions (Bonavoglia et al., 2006). In the first region, the gas sorbs without significant dilation of the polymer, in the second one, sorption occurs together with significant swelling, while in the last region swelling becomes negligible again. The same method could in principle be applied also to coal, where the sorbed concentration is replaced by the absolute adsorbed amount, which takes into account for both adsorption on the coal surface and absorption into the coal matrix. Both sorption and swelling isotherms can be described by a Langmuir equation as explained in Chapter 3 for the sorption experiments and shown in Figure 4.3 and 4.4 for the swelling data. The combination of the fitted sorption $n^t(P)$ and swelling $s(P)$ Langmuir isotherms, yields a relationship between swelling $s$ and amount adsorbed and sorbed $n^t$:

$$s_i = \frac{s_i^{max} b_{s,i} n_i^t}{b_{p,i} n_i^{max} - (b_{p,i} - b_{s,i}) n_i^t} \tag{4.6}$$

where $s_i^{max}$ and $b_{s,i}$ are the Langmuir coefficients for the swelling isotherms and $b_{p,i}$ and $n_i^{max}$ those for the sorption isotherm. Note that Eq.(4.6) is valid for $0 \leq n^t \leq n_i^{max}$. All the estimated values of the Langmuir constants for the coals tested here are summarized in Table 4.5. The outcome of this procedure is shown in Figure 4.5, where the volumetric swelling of coal samples I1 and I2 is shown as a function of the corresponding actual amount adsorbed and absorbed $n^t$. For the sake of clear representation, both swelling and sorption are normalized with the Langmuir constant $s^{max}$ and $n^{max}$, respectively. In both cases and for all the fluids, the obtained curves are similar: swelling increases monotonically with sorption. When compared to polymers, this behavior corresponds to the first part of the S-shape curve described above, where the appearance of this region is explained by the presence of void spaces, which are often encountered in polymers with a large degree of crystallinity (Bonavoglia et al., 2006). These spaces are the first to be occupied during the sorption experiment and as a consequence the bulk volume changes, i.e those measured using the view cell, are small. Coal is a porous material and the pores present in the coal matrix may therefore act similarly to the holes in a matrix of a crystalline polymer, thus explaining the behavior observed in this work. These results are also in good agreement with the literature, where a similar behavior was observed for a Illinois high volatile bituminous C coal (Day et al., 2008b) and for three Canadian coal of similar rank (Bustin et al., 2008).

## 4.3 Concluding remarks

Two coal samples obtained from two different coal mines in Italy have been investigated in terms of their volumetric behavior using a visualization technique. Experiments were carried out at condition relevant to ECBM, i.e. at 45°C and up to 130 bar. It was shown that exposing

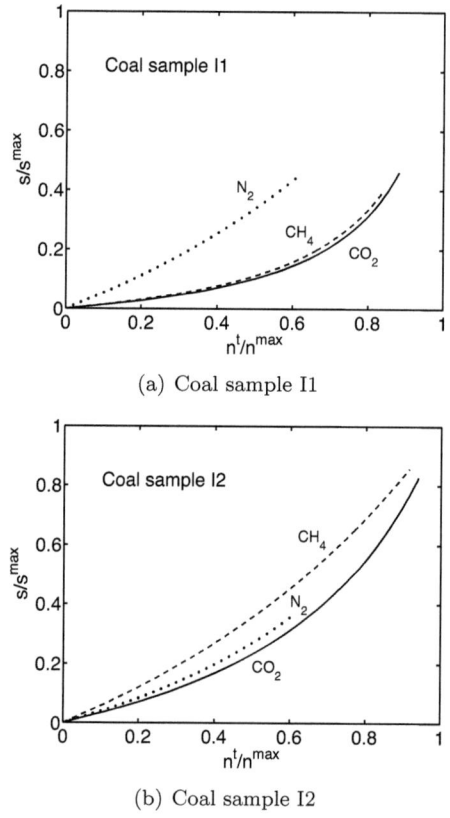

(a) Coal sample I1

(b) Coal sample I2

Figure 4.5: Dimensionless volumetric swelling $s/s^{\max}$ for coal samples (a) I1 and (b) I2 at 45°C as a function of dimensionless adsorbed and absorbed amount $n^t/n^{\max}$ for $CO_2$, $CH_4$ and $N_2$.

## 4.3 Concluding remarks

the coal to an adsorbing gas indeed leads to an expansion of the coal, with $CO_2$ showing the largest effect compared to $CH_4$ and $N_2$. No irreversibility of this phenomenon was observed, since after upon releasing the pressure the sample returned to its original size. The experimental results have been successfully described by a Langmuir-like equation.

## 4.4 Nomenclature

| | |
|---|---|
| $b_s$ | Langmuir equilibrium constant (swelling isotherm) [Pa$^{-1}$] |
| $b_p$ | Langmuir equilibrium constant (adsorption isotherm) [Pa$^{-1}$] |
| $d$ | Diameter of the coal sample [cm] |
| $d_0$ | Initial diameter of the coal sample [cm] |
| $h$ | Thickness of the coal sample [cm] |
| $h_0$ | Initial thickness of the coal sample [cm] |
| $n^a$ | Amount adsorbed and sorbed [mmol/g] |
| $n^{max}$ | Saturation capacity (adsorption isotherm) [mmol/g] |
| $P$ | Pressure [bar] |
| $R_o$ | Vitrinite Reflectance [%] |
| $\rho^b$ | Bulk density [g/cm$^3$] |
| $s$ | Swelling [-] |
| $s^{max}$ | Saturation capacity (swelling isotherm) [-] |
| $T$ | Temperature [°C] |
| $V^{coal}$ | Volume of the coal sample [cm$^3$] |
| $V_0$ | Initial volume of the coal sample [cm$^3$] |
| $\Delta V^s$ | Volume of the sorbed phase [cm$^3$] |

# Chapter 5

# Permeability of coal

## 5.1 Introduction

[1]Volume changes of coal during ECBM operations are of key importance because they affect coal bed permeability, which in turn controls injection pressure and gas production. During the performed ECBM field tests (see Chapter 1), operators were forced to reduce injection rates due to an unexpected pressure increase at the injection well, which was attributed to a decrease in permeability caused probably by swelling of the coal (Van Bergen et al., 2006; Wong et al., 2007; Yamaguchi et al., 2006). Being at several hundreds meter of depth, the coal seam is subjected to overburden and lateral stresses. Both fluid pressure and volumetric strain (swelling/shrinkage) induced by gas adsorption/desorption induce changes in the stress field of the coal seam. Fractures (cleats)

---
[1]This work has been published as Pini et al. (2009).

undertake most of the deformation upon stress changes, being very sensitive to them as compared to the coal matrix (Cui et al., 2007). A variation in the cleats opening is definitively affecting the permeability of the coal seam, which is the main petrophysical property controlling the performance of the ECBM operation. This phenomenon needs to be quantified, since, besides affecting $CO_2$ injectivity and $CH_4$ recovery, it hinders an optimal exploitation of the coal seam.

Beside the measurement of single and multicomponent gas sorption on coal, experiments aimed at the study of the swelling phenomenon and its consequences on the coal permeability are just beginning. Recently, ECBM recovery by gas injection has been investigated by injecting $CO_2$ into a chromatographic column packed with crushed coal (Tang et al., 2005; Yu et al., 2008a). Although these experiments allowed to obtain insights on the $CO_2/CH_4$ displacement mechanism, they are not suited to study the gas flow properties which depend on the natural fractured structure of the coal, which is lost with the grinding. The field examples given above, suggest that this structure indeed controls the dynamics of gas displacement and therefore experiments should be carried out with coal cores under in situ conditions (confinement), i.e. under conditions similar to those which can be found deep underground. Harpalani and coworkers (Harpalani and Schraufnagel, 1990) measured changes in permeability of coal cores confined by a constant hydrostatic pressure as a consequence of $CH_4$ desorption: results showed that permeability decreases with decreasing gas pressure, because of the narrowing of the cleats caused by the increasing effective stress on the sample. It was further observed, that if an adsorbing gas is used, the matrix shrinks at low pressures when significant desorption starts: this phenomenon counteracts the narrowing of the cleats resulting in some cases in a net permeability increase (Harpalani and Chen, 1997). More recently, the

## 5.1 Introduction

effect of $CO_2$ injection on the permeability of coal samples was investigated with a high-pressure core flooding setup (Mazumder et al., 2006a; Mazumder and Wolf, 2008). The experiments were carried out at 45°C by imposing a constant effective stress on the sample. Injection of $CO_2$ resulted in volumetric expansion (swelling) of the sample and the observed increase in permeability with $CO_2$ pressure was attributed to the fact that the coal sample was allowed to expand freely inside the pressure cell (Mazumder et al., 2006a). Moreover, differential swelling was obtained by measuring the axial changes of the core dimension during the injection of $CO_2$ in a coal core, which has been pre-saturated with $CH_4$, and its effects on the core permeability were estimated (Mazumder and Wolf, 2008). Finally, permeability core flood tests were performed at 25°C on cubic coal samples using a triaxial stress coal permeameter (Massarotto et al., 2007; Wang et al., 2007). The volumetric changes of the coal sample caused by gas adsorption were obtained by three dimensional strain measurements and a dramatic permeability drop was observed during the $CO_2/CH_4$ displacement experiment. Moreover, rigorous modeling was used to describe the dynamics of the process (Wang et al., 2007).

In the present chapter, an experimental technique is presented and results shown from flow experiments carried out under coal seam conditions (45°C and different confining and gas reservoir pressures) by injecting different gases (He, $CO_2$ and $N_2$) in a dry coal core subject to several levels of hydrostatic confinement. The technique used is the transient step method, which is different from the one used in the studies mentioned above, i.e. the constant pressure difference method (Fisher, 1992). The dynamic behavior observed during the experiments is described by a mathematical model including mass balances accounting for gas flow, adsorption and swelling, and mechanical constitutive equations for the description of porosity and permeability changes during injection. The combination between such a model and the dynamic experiments mentioned above represents the novelty of this work. By predicting the gas flow experiments under different conditions of effective stress, this approach enables the evaluation of key quantities controlling the performance of the ECBM process. In particular, the obtained relationship between permeability and operating pressures can be directly implemented in the models typically used to simulate reservoir dynamics and to history match field test data obtained during ECBM operations.

## 5.2 Experimental section

### 5.2.1 Coal sample characterization

A coal sample from the Monte Sinni coal mine in the Sulcis Coal Province (Sardinia, Italy) was used. The sample was drilled in December 2004 at a depth of about 500 m and preserved in a plastic box in air. Results of a thermo-gravimetric analysis (TGA) give a coal composition of 49.4 % in fixed carbon content, 41.2 % in volatile matter, 2.1 % in ash and 7.3 % in moisture. These values, together with a vitrinite reflectance coefficient ($R_o \cong 0.7$), allow classifying the coal as high volatile C bituminous (Ottiger et al., 2006).

For the flow measurements reported in this study, a single coal core of 2.54 cm (1 inch) in diameter and 3.6 cm in length obtained from the sample above was used. Since the coal is brittle, the drilling of the core with a hollow diamond drill and using water as a cooling medium was unsuccessful. A new procedure was therefore applied for its preparation. Firstly, the coal core was roughly pre-shaped on a steel bandsaw. Secondly, after setting the coal core in a self designed holder, it was gently ground on a

## 5.2 Experimental section

band grinder to the desired diameter, that is 2.54 cm ± 0.01 cm. In the final step, the planar ends of the cylindrical sample were polished. The sample was vacuum dried in a oven at 70°C for at least 2 days before the experiments.

To determine the coal sample porosity, helium density and mercury porosimetry measurements were performed using a Helium Pycnometer 1330 (Micromeritics Instrument, Belgium) and a Pascal 440 Porosimeter (Thermo Electron Corporation, Germany), respectively. Coal possesses a complex porous structure which is usually characterized by cleats, i.e. the fracture system allowing for gas flow, macropores, where gas is present but does not flow by convection, and micropores, where adsorption takes place (see Chapter 2). If the microporosity is accounted for as combined with the solid material, the total coal sample porosity $\varepsilon^*$ can be defined as

$$\varepsilon^* = \varepsilon + (1-\varepsilon)\varepsilon_\mathrm{p} \tag{5.1}$$

where $\varepsilon$ and $\varepsilon_\mathrm{p}$ are the cleat and macropore porosity, respectively. Macropore porosity of a powdered coal sample from the same batch as the coal used in the present study has been determined from helium and mercury density measurement as $\varepsilon_\mathrm{p} = \rho_{\mathrm{Hg}}(1/\rho_{\mathrm{Hg}} - 1/\rho_{\mathrm{He}})$. The total interconnected porosity of the unstressed core sample $\varepsilon_0^*$, which includes macropores and cleats, was estimated from the bulk volume $V_\mathrm{b}$ (calculated from the radius and length of the specimen) and the skeletal volume of the adsorbent $V_\mathrm{ads}$ (measured using the helium pycnometer) as $\varepsilon_0^* = 1 - V_\mathrm{ads}/V_\mathrm{b}$. With application of Eq. (5.1), the initial cleat porosity $\varepsilon_0$ can now be estimated. Properties of the coal sample used in this study are given in Table 5.1. Note that in the course of the gas flow experiments (Section 5.3) cleat porosity changes depending on the specific stress situation of the coal sample and as a consequence the total porosity changes as well.

Table 5.1: Model input parameters

| Property | Value |
|---|---|
| $T$ [K] | 318.15 |
| $\varepsilon_0^*$ [-] | 0.051 |
| $\varepsilon_p$ [-] | 0.020 |
| $\varepsilon_0$ [-] | 0.032 |
| $\nu$ [-] | 0.26 |
| $E_Y$ [Pa] | $1.119 \times 10^9$ |
| $\rho_{ads}$ [kg/m$^3$] | 1356.6 |
| $L$ [m] | 0.036 |
| $A$ [m$^2$] | $4.73 \times 10^{-4}$ |
| $V_{US}$ [m$^3$] | $5.04 \times 10^{-5}$ |
| $V_{DS}$ [m$^3$] | $1.52 \times 10^{-5}$ |

Sorption and swelling isotherms were measured on coal samples from the same batch as the coal used in the present study (see Chapter 2 and 4). $CO_2$ and $N_2$ sorption isotherms on coal were obtained using a Magnetic Suspension Balance (Rubotherm, Germany) (Ottiger et al., 2008a). Prior to the adsorption experiments, the coal sample was ground to a size between 250 and 355 $\mu$m and dried in an oven at 105°C under vacuum for one day. In order to be used in the model presented in this work, the measured excess sorption isotherms have been converted to absolute adsorbed amounts by assuming a constant adsorbed phase density with values of $\rho_{CO_2}^a = 36.7$ mol/L and $\rho_{N_2}^a = 47.1$ mol/L (Sudibandriyo et al., 2003b; Ottiger et al., 2008b).

Swelling isotherms of $CO_2$ and $N_2$ were obtained using a high pressure view cell equipped with a digital photocamera (Ottiger et al., 2008a). A coal disc was drilled from a coal block and prior to the swelling experiments, it was dried in an oven at 70°C under vacuum for two days.

## 5.2 Experimental section

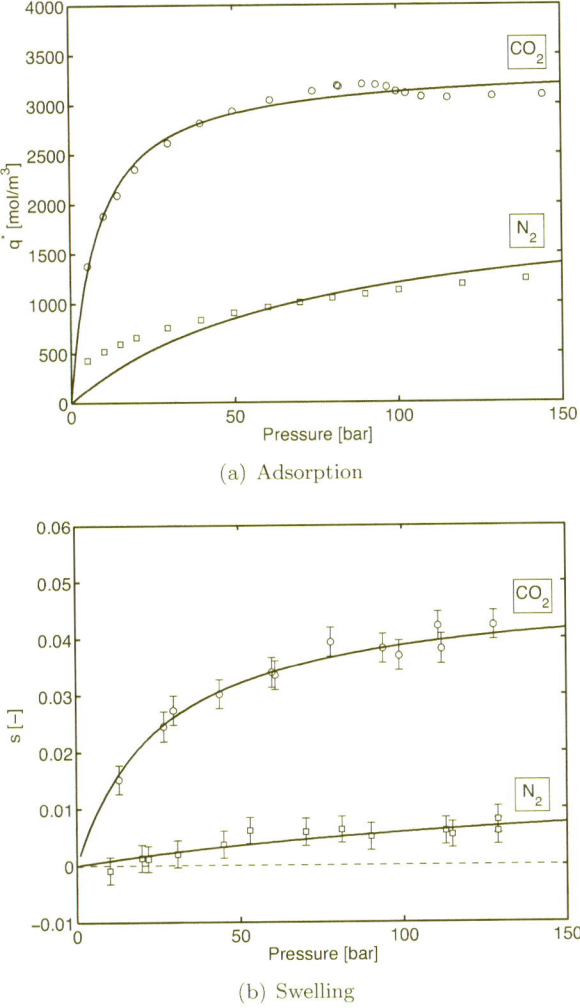

Figure 5.1: (a) Adsorption and (b) swelling isotherms at 45°C as a function of pressure for $CO_2$ (O) and $N_2$ (□) measured on a coal sample from the Sulcis coal province. Solid lines correspond to the Langmuir model.

The measured $CO_2$ and $N_2$ adsorption and swelling isotherms have been fitted by the Langmuir model and they are shown in Figure 5.1. All the estimated values of the Langmuir constants are summarized in Table 5.2.

The pure gases used in this study were obtained from Pangas (Luzern, Switzerland), namely $CO_2$ at a purity of 99.995 % and $N_2$ and He at purities of 99.999 %.

Table 5.2: Langmuir isotherm parameters

| | $q^*(c) = \frac{q_{m,i} b_i c}{1+b_i c}$ | | $q^*(P) = \frac{q^p_{m,i} b^p_i P}{1+b^p_i P}$ | | $s(P) = \frac{s_{m,i} b^s_i P}{1+b^s_i P}$ | |
|---|---|---|---|---|---|---|
| **Species** | $q_{m,i}$ | $b_i$ | $q^p_{m,i}$ | $b^p_i$ | $s_{m,i}$ | $b^{s,i}$ |
| | [mol/g] | [m$^3$/mol] | [mol/g] | [1/Pa] | [-] | [1/Pa] |
| $CO_2$ | $2.38 \times 10^{-3}$ | $3.55 \times 10^{-3}$ | $2.49 \times 10^{-3}$ | $1.25 \times 10^{-6}$ | $4.90 \times 10^{-2}$ | $3.80 \times 10^{-7}$ |
| $N_2$ | $1.06 \times 10^{-3}$ | $1.11 \times 10^{-3}$ | $1.52 \times 10^{-3}$ | $1.40 \times 10^{-7}$ | $1.70 \times 10^{-2}$ | $5.19 \times 10^{-8}$ |

## 5.2.2 Experimental Setup

The experimental set-up to carry out the flow experiments is shown in Figure 5.2 and was mostly developed and built in-house. The heart of the set-up is a hydrostatic cell, designed to accommodate cylindrical samples of about 2.5 cm in diameter and up to 5 cm in length, and to work with confining pressures up to 1000 bar. The hydrostatic cell is kept at the desired temperature with a heating jacket thermostated using a liquid thermostat Huber Unistat T325 (Renggli, Rotkreuz, Switzerland). Temperature is measured with a K-type thermocouple with an accuracy of 0.1°C placed close to the sample. At the experimental temperature used in this work, temperature gradients along the sample were found to be smaller than the accuracy of the temperature measurement. The con-

## 5.2 Experimental section

fining pressure is measured with a calibrated pressure sensor Intersonde (Barnbrook Systems Ltd., Hampshire, UK) and it is controlled ($\pm$ 1 bar) with a screw type displacement pump driven by a microstepping motor (NovaSwiss, Effretikon, Switzerland). As a confining fluid, hydraulic oil was used (Morlina Oil, ISO VG 10, Shell) and beside the displacement pump, the confining pressure system consists of a hand pump for initial pressure build up (NovaSwiss, Effretikon, Switzerland) and a circulation pump (Verder, Haan, Germany). The cylindrical sample is isolated from the confining fluid with a 4 mm thick polyvinyl chloride rubber jacket and is placed between two stainless steel disks with interconnected circular grooves to distribute the fluid over the cross-sectional area of the sample. The two stainless steel disks are connected to the tubing system and finally to two reservoirs: the upstream reservoir, which can be pressurized with the gas to be injected, and the downstream reservoir used to collect the gas which leaves the sample. Calibration of the reservoirs' volumes gave values of 50.4 cm$^3$ and 15.2 cm$^3$, respectively. The reservoirs are placed in a water bath which is maintained at the same temperature as the hydrostatic cell with a Haake N3 thermostat (Digitana AG, Horgen, Switzerland). The pressures in the two reservoirs is measured with two pressure sensors with an accuracy of 0.05 % (PAA-35HTT, Keller, Winterthur, Switzerland).

### 5.2.3 Measurement procedure

The transient step method was used to carry out the flow experiments. This technique has been widely used to measure the permeability of rocks, in particular of low permeability rocks, due to the advantage of measuring pressures instead of flow rates in a high pressure experiment (Brace et al., 1968; Fisher, 1992; Hildenbrand et al., 2002). In a typical

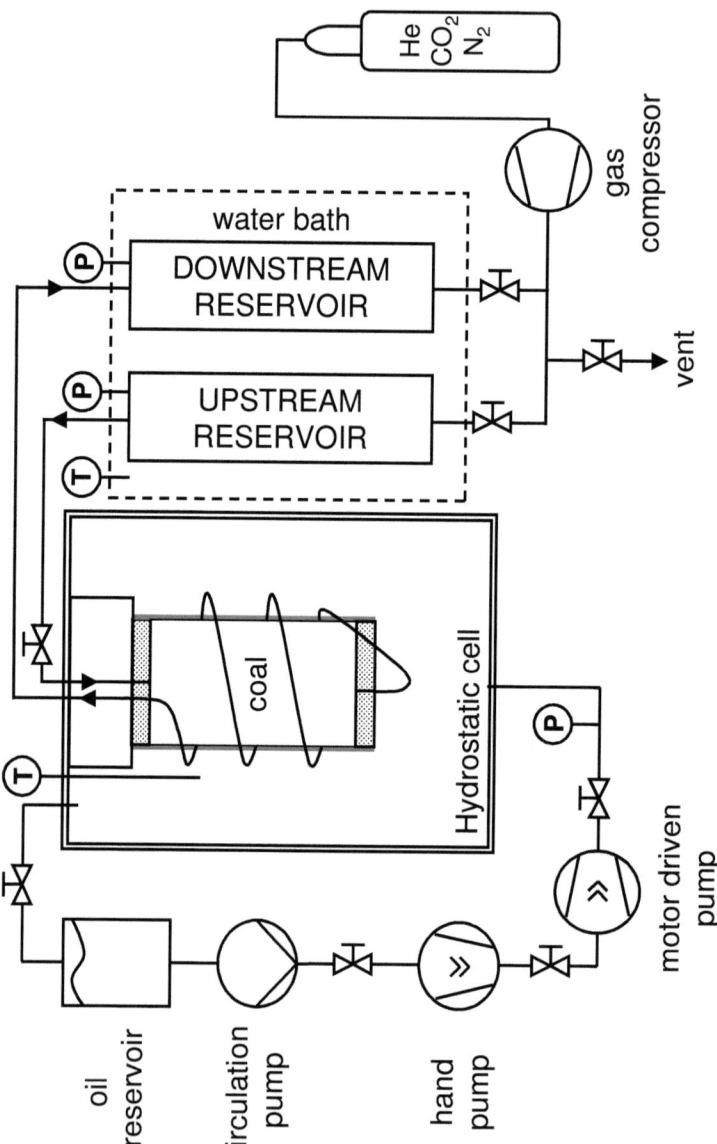

Figure 5.2: Setup used in this study for the permeability measurements under confined conditions.

## 5.2 Experimental section

experiment, the assembled sample is placed into the hydrostatic cell and a confining pressure is applied and held constant. The sample is then flushed with the fluid to be used and, as an initial condition, reservoirs and sample are equilibrated with a fluid at the same pressure. Moreover, when a new gas is used the flushing is repeated several times at a higher pressure to remove residual traces of the previous gas. Before starting the experiment, two days are allowed to establish uniform adsorption equilibrium conditions in the coal core. A pressure change is then imposed at the upstream end of the sample and the system is allowed to equilibrate at a new pressure level. After reaching pressure equilibrium and allowing for at least two days for the fluid to adsorb, the pressure in the upstream reservoir is risen again to a new level, and a new measurement is carried out. In doing so, a wide range of conditions can be tested in terms of effective pressure on the sample (defined as the difference between confining and pore pressure, $P_e = P_c - P$). In this study, the confining pressure ranged from 60 to 140 bar and gas reservoir pressures varied from 10 to 80 bar. These values were chosen to cover the range of conditions representative for the Sulcis coal seam at 500 m depth. More precisely, the confining pressure was chosen to be close to the lithostatic overburden stress $T_{zz}$, which is a function of the rock mass density $\gamma$ above the coal seam and of the depth of the seam $z$, i.e. $T_{zz} = \int_z \gamma dz$ with $\gamma$ varying between 0.16 - 0.20 bar/m (Atkinson, 2007). The injection gas pressures were taken to be similar to those used in previous field studies (Wong et al., 2007; Yamaguchi et al., 2006).

The observed pressure decay in the upstream reservoir and the pressure rise at the downstream side are then compared to the behavior predicted by a suitable model. In particular, two types of experiments are carried out: those where the confining pressure is kept constant and the gas reservoir pressure is raised, and those where the same pressure step is repeated under different constant levels of confining pressure. Different fluids have been used in the experiments, namely He, $N_2$ and $CO_2$, and as explained in the Section 5.4, the experiments with the non-adsorbing helium are needed to study the effect of the confining pressure on the sample permeability, whereas those with adsorbing fluids ($N_2$ and $CO_2$) to study the effects of adsorption and swelling on the flow dynamics.

## 5.3 Modeling

The experimental procedure described above does not allow for a direct measurement of the permeability of the coal sample. In order to obtain this quantity the observed behavior of the pressure in the two reservoirs is compared to the behavior predicted theoretically, i.e. by applying a model that describes the dynamics of the system upstream reservoir-coal core-downstream reservoir. In its simplest form under steady state conditions such a model would consist only of Darcy's law, which describes the flow of a fluid through a porous medium (Hildenbrand et al., 2002). However, coal can adsorb several gases and that as a consequence it expands. Moreover, stress changes caused by the difference in pore and confining hydrostatic pressure definitely affect the porosity and permeability of the sample. Therefore, in the development of a model describing the unsteady flow of gases through a coal sample all these phenomena have to be taken into account.

A so-called core model has been written; it consists of one-dimensional mass balances combined with stress-strain relationships, where the mechanical behavior of the coal sample is described. The coal core is characterized by the fracture network (cleats), which divides the core into different matrix blocks (Gilman and Beckie, 2000). The fracture system, with its higher permeability, allows for the flow through the core, whereas the relatively impermeable porous matrix is instead mainly responsible for gas storage, in terms of free gas (macropores) and sorption (micropores).

### 5.3.1 Mass Balances

All the experiments carried out in this study have been performed using pure gases (He, $CO_2$ or $N_2$) and therefore mass balances are written and solved for one component only. However, for the sake of clarity an index $i$ is added to the parameters which are component specific. The overall mass balance equation for the fluid in the coal core takes the following form:

$$\frac{\partial(\varepsilon^* c)}{\partial t} + \frac{\partial[(1-\varepsilon^*)q]}{\partial t} + \frac{\partial(uc)}{\partial z} = 0 \qquad (5.2)$$

where $c$ and $q$ are the gas and adsorbed phase concentrations, respectively, $\varepsilon^*$ is the total interconnected porosity, $u$ is the superficial velocity, $t$ is the time and $z$ the axial coordinate. By simulating the rate of diffusion through the coal matrix by a linear driving force model, the material balance for the adsorbed phase can be written as follows:

$$\frac{\partial[(1-\varepsilon^*)q]}{\partial t} = (1-\varepsilon^*)k_{M,i}(q^* - q) \qquad (5.3)$$

where $k_{M,i}$ is the mass transfer coefficient of component $i$ and $q^*$ is the adsorbed gas concentration in equilibrium with the gas phase. Note that since Helium is considered to be an inert, both adsorption on coal and mass transfer in the adsorbed phase are absent, and therefore Eq.(5.3) is considered for $N_2$ and $CO_2$ only.

The superficial velocity $u$ is related to the pressure gradient through the Darcy's equation:

$$u = v\varepsilon = -\frac{k}{\mu}\left[\frac{\partial P}{\partial z} - gM_{m,i}\left(z\frac{\partial c}{\partial z} + c\right)\right] \quad (5.4)$$

where $v$ is the interstitial velocity and $\varepsilon$ the cleat porosity, $P$ the gas pressure, $k$ the permeability, $\mu$ the dynamic viscosity, $g$ the gravitational acceleration and $M_{m,i}$ the molecular weight of the fluid ($i=$ He, $CO_2$ or $N_2$). Note that the second term in square brackets accounts for the gravitational effects, being the sample positioned vertically in the hydrostatic flow cell. Data from the National Institute of Standards and Technology (NIST) have been interpolated with polynomials in the temperature and pressure range used in this study to relate pressure and viscosity to the fluid concentration (NIST, 2008). As anticipated, the adsorption on coal is described by the Langmuir isotherm as a function of the fluid concentration $c$:

$$q^*(c) = \frac{\rho_{ads} q_{m,i} b_i c}{1 + b_i c} \quad (5.5)$$

where $\rho_{ads}$ is the coal bulk density and $q_{m,i}$ and $b_i$ are the saturation capacity per unit mass adsorbent and the Langmuir equilibrium constant, respectively. Finally, the mass balances are completed by giving the equations for the upstream and downstream reservoirs:

$$\left(\frac{\partial c}{\partial t}\right)_{z=0} = -\frac{A}{V_{US}}(uc)_{z=0}$$
$$\left(\frac{\partial c}{\partial t}\right)_{z=L} = \frac{A}{V_{DS}}(uc)_{z=L} \quad (5.6)$$

where $A$ and $L$ are cross-sectional area and length of the sample, and $V_{US}$ and $V_{DS}$ are the volumes of the upstream and downstream reservoirs, respectively.

### 5.3.2 Stress-Strain Relationship

Throughout the experiment, changes in the fluid pressure and the swelling caused by adsorption induce changes in the coal sample stress field. As a consequence, porosity and permeability will also change. Typically, the variation in coal permeability with respect to an arbitrary reference state 0 is related to the change in cleat porosity by the following equation (Palmer and Mansoori, 1998; Cui et al., 2007; Wei et al., 2007a):

$$\left(\frac{k}{k_0}\right) = \left(\frac{\varepsilon}{\varepsilon_0}\right)^3 \tag{5.7}$$

This relationship between porosity and permeability is very popular and well accepted, especially in the case of coals exhibiting fairly regular structure (matchstick geometry) (Reiss, 1980; Seidle et al., 1992; Harpalani and Chen, 1997; Pekot and Reeves, 2003). The description of the mechanical behavior of the coal sample can be effectively achieved by the stress-strain constitutive equations of an isotropic linear poroelastic medium (Palmer and Mansoori, 1998; Gilman and Beckie, 2000; Shi and Durucan, 2004a; Cui et al., 2007; Zhu et al., 2007). The isotropic simplification is justified since no macroscopic preferred orientation of heterogeneities and microcracks was observed. Therefore, only two parameters are sufficient to describe the elastic behavior of the coal (here $E_Y$ and $\nu$). The stress ($T_{jk}$) and strain ($E_{jk}$) relation for the coal sample can be written as follows, where $j, k$ can be any combination of $x$, $y$ and $z$ (Neuzil, 2003):

$$T_{jk} = \frac{E_Y}{1+\nu}\left(E_{jk} + \frac{\nu}{1-2\nu}e\delta_{jk}\right) + P\delta_{jk} + Ks\delta_{jk} \quad (5.8)$$

where $E_Y$ and $\nu$ are the coal sample Young's modulus and Poisson's ratio, respectively, $e$ is the bulk volumetric strain ($e = E_{xx} + E_{yy} + E_{zz}$), $K$ is the bulk modulus ($K = E_Y/[3(1-2\nu)]$), $s$ the swelling and $\delta_{jk}$ the Kronecker's delta. Note that the term accounting for swelling on the right-hand side of Eq.(5.8) is analogous to the thermal expansion term in the non-isothermal expansion of a poroelastic solid. In particular here the approach presented by Cui et al. (2007) is followed, where a general solution is found for the porosity change $\varepsilon/\varepsilon_0$ as a function of the mean effective stress change, $\Delta T_e$, i.e.

$$\frac{\varepsilon}{\varepsilon_0} = \exp\left(\frac{(T-T_0)-(P-P_0)}{K\varepsilon_0}\right) = \exp\left(\frac{-\Delta T_e}{K\varepsilon_0}\right) \quad (5.9)$$

where $T = (T_{xx}+T_{yy}+T_{zz})/3$ is the mean normal stress, $T_e = T - P$ the mean effective stress and subscript 0 denotes a reference state. Starting from the stress-strain constitutive equations, i.e. Eq.(5.8), a general relationship for the change in effective mean stress $\Delta T_e$ is found, which takes the following form:

$$\Delta T_e = -C_p(P-P_0) + C_{s,i}E_Y(s-s_0) + C_o \quad (5.10)$$

where $C_p$ and the $C_{s,i}$ are respectively the coefficients accounting for changes in the porosity due to gas pressure and swelling, and $C_o$ is a constant associated with the boundary conditions (confinement). Note

## 5.3 Modeling

that $C_o$, $C_{s,i}$ and $C_p$ are always positive. During the experiments, the coal core was always hydrostatically loaded by a constant confining pressure and therefore the constant $C_o$ can be written as the product of a dimensionless constant and the confining pressure, $C_o = C_c P_c$. Moreover, due to the hydrostatic conditions, which hold both for the pore pressure (homogeneous fluid distribution in the sample) and the confining pressure, the effects of gas pressure and confining pressure on the porosity can be assumed to be equal, i.e. $C_c = C_p = C_e$. Finally, by taking as a reference state the one without confinement and gas pressure ($P_c = P_0 = s_0 = 0$), the following equation is obtained:

$$\frac{\varepsilon}{\varepsilon_0} = \left(\frac{k}{k_0}\right)^{1/3} = \exp\left(\frac{-C_e(P_c - P) - C_{s,i} E_Y s}{K \varepsilon_0}\right) \qquad (5.11)$$

Note that the underlying assumption in Eq.(5.11) is $P_c \geq P$, reflecting the practical constraint of confining pressure larger or equal to gas pressure. It is worth noting that, in agreement with other models reported in the literature, also in this case the porosity and permeability changes are defined by two terms, one accounting for the effects caused by the pressure and a second one for those caused by swelling (Levine, 1996; Palmer and Mansoori, 1998; Gilman and Beckie, 2000; Shi and Durucan, 2004a). Eq.(5.11) is completed by giving a relationship for the swelling $s$. Combining the fitted adsorption $q^*(P)$ and swelling $s(P)$ Langmuir isotherms (see Chapter 4), yields a relationship between swelling and amount adsorbed:

$$s_i = \frac{s_{m,i} b_i^s q}{b_i^p q_{m,i}^p \rho_{\text{ads}} - (b_i^p - b_i^s) q} \qquad (5.12)$$

where $s_{m,i}$ and $b_i^s$ are the Langmuir coefficients for the swelling isotherms and $b_i^p$ and $q_{m,i}^p$ those for the adsorption isotherm. Note that Eq.(5.12) is valid for $0 \leq q \leq q_{m,i}^P$. The advantage of giving the swelling as a function of the adsorbed amount instead of the pressure, is that in the former case the kinetic of the swelling process is accounted for through the adsorption rate given by Eq.(5.3), whereas in the latter case this would not be the case, being the pressure equilibration instantaneous.

### 5.3.3 Solution Procedure

The problem is defined by Eqs. (5.2)-(5.7), (5.11) and (5.12). The orthogonal collocation method has been applied to discretize in space the partial differential equations (Villadsen and Michelsen, 1978; Morbidelli et al., 1983). The resulting system of ordinary differential equations has then been solved numerically using a commercial ODEs solver (in Fortran). The input parameters used for the model calculations are summarized in Tables 5.1 and 5.2. All the parameters are specific for the coal sample used in this study, except for the Poisson's ratio $\nu$ and the Young's modulus $E_Y$, whose values have been taken from the literature for a coal similar to our own (Gentzis et al., 2007). The set of parameters to be estimated consists of the coefficients $C_e$ and the $C_{s,i}$ accounting respectively for changes in the porosity due to effective pressure and swelling, the absolute (unstressed sample) permeability $k_0$ and the mass transfer coefficients $k_{M,i}$. All parameters were estimated by reproducing the experimentally obtained transient steps using the model described above. In particular, the experiments with helium, i.e. the non-adsorbing (and non-swelling) gas, were used to obtain values for $C_e$ and $k_0$ by minimizing the following objective function $\Phi$ using a Simplex algorithm:

$$\Phi = \sum_{j=1}^{N_{\text{exp}}} \left[ \left( \frac{P_{\text{US},j}^{\text{exp}} - P_{\text{US},j}^{\text{mod}}}{P_{\text{US},j}^{\text{exp}}} \right)^2 + \left( \frac{P_{\text{DS},j}^{\text{exp}} - P_{\text{DS},j}^{\text{mod}}}{P_{\text{DS},j}^{\text{exp}}} \right)^2 \right] \quad (5.13)$$

where $N_{\text{exp}}$ is the number of experimental data points, $P$ the pressure in the upstream (US) and downstream (DS) reservoir, and the superscripts exp and mod refer to the variables obtained from the experiments and from the model, respectively. The experiments with the adsorbing gases, i.e. $CO_2$ and $N_2$, were then used to estimate the values for the coefficients $C_{s,i}$ and those of the mass transfer coefficients $k_{M,i}$, using the values of $C_e$ and $k_0$ determined with the helium experiments again by minimizing the objective function $\Phi$ defined by Eq.(5.13).

## 5.4 Results and Discussion

In this section, the results of the transient step experiments carried out with He, $N_2$ and $CO_2$ are presented. All the experiments were carried out at 45°C, a temperature which is representative of the conditions of the coal seam in the Sulcis Coal Province in Sardinia, Italy. In order to compare all the obtained data, each single transient step is analyzed as shown in Figure 5.3: four specific points characterize the transient step, namely the initial ($P_{\text{in}}$) and final ($P_{\text{eq}}$) gas reservoir pressures in the downstream reservoir, the confining pressure ($P_c$) and the time needed to complete 50 % of the imposed pressure step ($\tau_{0.5}$), i.e. the time to reach $P_{0.5} = P_{\text{in}} + 0.5(P_{\text{eq}} - P_{\text{in}})$.

In Figure 5.4, the obtained $\tau_{0.5}$ are shown together with those predicted by the model as a function of the effective pressure on the sample, i.e. $P_e = P_c - P_{\text{eq}}$. Note that, following this approach, all the experimental points can be compared, since for each transient step a similar pressure

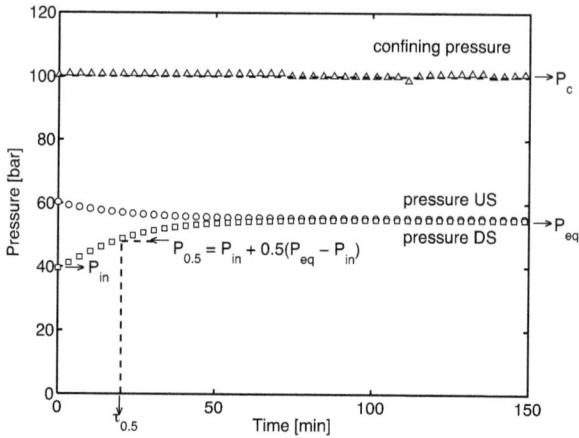

Figure 5.3: Example of an experimental transient step: confining pressure $P_c$ (△), upstream $P_{US}$ (○) and downstream $P_{DS}$ (□) reservoir pressures as a function of time.

increase of about 20 bar was imposed. With the exception of some helium points obtained at the lowest effective pressure $P_e$, it can be seen that over the whole range of effective pressures tested, the experimental reproducibility is satisfactory, being the time scale measured very large. It is worth noting that the relatively low maximal confining pressure could ensure an elastic behavior of the coal sample. This observation is supported by the reproducibility of the experimental points shown in Figure 5.4 even if we mixed up conditions and components when performing the experiments. A linear dependency of $\tau_{0.5}$ on the effective pressure $P_e$ is observed, which is in agreement with the model predictions (dashed lines): for all the three gases, $\tau_{0.5}$ is increasing with $P_e$, due to the compaction of the sample and the consequent narrowing of the cleats apertures under an increasing imposed stress. In addition, for a specific effective pressure, the time to reach equilibration is different for the three

## 5.4 Results and Discussion

fluids, with Helium being faster than $N_2$, and $N_2$ faster than $CO_2$. As it will be explained with more details below, in contrast to the inert Helium, the adsorption and the consequent volumetric expansion of the coal sample when exposed to an adsorbing gas is in fact affecting the fluid flow through the coal.

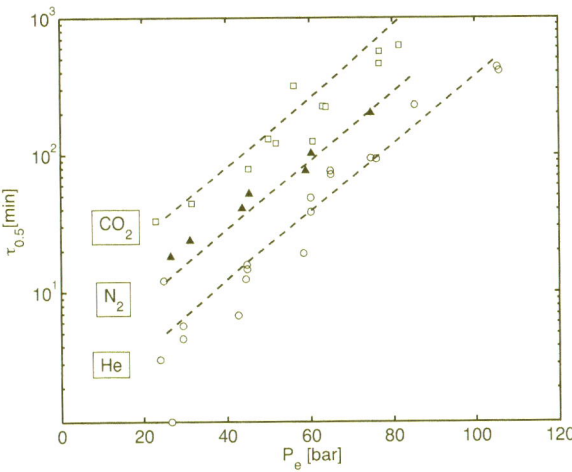

Figure 5.4: Time to complete 50% of the imposed transient step, $\tau_{0.5}$ (logarithmic scale) as a function of the effective pressure $P_e = P_c - P_{eq}$ on the sample when Helium (○), $N_2$ (▲) and $CO_2$ (□) were injected. Dashed lines represent model results.

### 5.4.1 Experiments with an Inert Gas

In this section the results obtained when Helium was used are presented and analyzed in greater detail. The experiments have been carried out at 45°C and various effective pressures, and they are used to estimate those parameters which are specific to the coal sample tested, namely the initial (unstressed) permeability $k_0$ and the effective pressure coefficient $C_e$. Being independent of the fluid used, these parameters can then be applied in the simulations where an adsorbing gas is injected instead of Helium. In Figure 5.5 two examples of transient steps are shown, which were carried out (a) by keeping the confining pressure constant at 100 bar and increasing the fluid pressure, and (b) by repeating the same pressure step under different confining pressures. The symbols are the experimental data whereas the solid lines correspond to the model results. The range of gas reservoir pressure tested was between 10 and 80 bar, whereas the confining pressure was changed between 60 and 140 bar. A good agreement is observed between experiments and simulated transient steps, which were obtained by fitting the initial (unstressed) permeability $k_0$ and the effective pressure coefficient $C_e$ to the experimental data. The fitted values are summarized in Table 5.3.

Table 5.3: Estimated values of the model fitting parameters

| Parameter | Species | | |
|---|---|---|---|
| | Helium | $N_2$ | $CO_2$ |
| $k_0$ [mD] | 0.049 | | |
| $C_e$ [-] | 4.676 | | |
| $C_{s,i}$ [-] | - | 2.377 | 0.622 |
| $k_{M,i}$ [s$^{-1}$] | - | 2.262×10$^{-6}$ | 3.878×10$^{-6}$ |

It is worth noting, that the obtained permeability $k_0$ (0.049 mD) is smaller than that of typical coal beds measured in the field, which ranges between 1 and 10 mD (White et al., 2005), but it is similar to the permeability values obtained in other laboratory studies (Harpalani and Schraufnagel, 1990; Harpalani and Chen, 1997; Mazumder et al., 2006a). This discrepancy between laboratory and results from the field can be attributed to the absence in the small samples of the large fractures which on the contrary are present in the coal seam and represent an

## 5.4 Results and Discussion

important contribution as far as the gas flow is concerned. Finally, the good agreement between experiments and model validates the selected relationship for permeability and porosity.

It can also be seen in Figure 5.5, that the transient step equilibration time increases with increasing effective pressure on the sample, i.e. $P_e = P_c - P_{eq}$. As given in Eqs. (5.7) and (5.11), the permeability and the porosity decrease with increasing effective pressure, thus slowing down the flow process. This phenomenon corresponds in practice to the compression of the cleats under an external stress. This effect is illustrated in Figure 5.6, where relative porosity $(\varepsilon/\varepsilon_0)$ and permeability $(k/k_0)$ curve of the coal sample are reported as a function of the effective pressure $P_e$. The curve has been obtained by solving Eqs.(5.7) and (5.11) for different values of effective pressure, and by setting the swelling $s$ to zero. The permeability and porosity data representing the end of each transient step, i.e. when the pressure in the downstream reservoir reaches $P_{eq}$, are also shown in Figure 5.6.

Let us note that such data have been calculated using the experimental values of $P_{eq}$ and $P_c$ in Eq.(5.11). The numerical values of the same data are also reported in Table 5.4 together with the corresponding confining and equilibrium downstream pressures. As expected, when the effective pressure reaches the value of zero, the original permeability and porosity are completely recovered.

### 5.4.2 Experiments with an Adsorbing Gas

In this section the results obtained when $CO_2$ and $N_2$ were injected into the coal core at 45°C and exposed at several confining pressures are presented. Since the coal sample specific parameters are already known from the Helium experiments, the injection experiments performed with

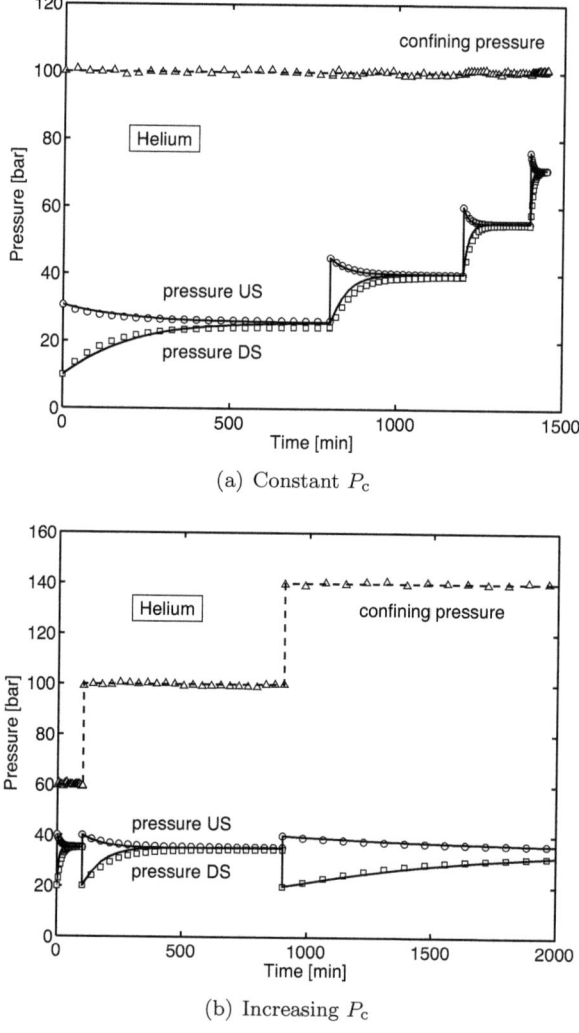

(a) Constant $P_c$

(b) Increasing $P_c$

Figure 5.5: Transient steps measurements at 45°C when Helium is injected: (a) confining pressure kept constant and (b) varying confining pressure. Confining pressure $P_c$ ($\triangle$), upstream $P_{US}$ ($\bigcirc$) and downstream $P_{DS}$ ($\square$) reservoir pressures as a function of time. Solid lines correspond to model results.

## 5.4 Results and Discussion

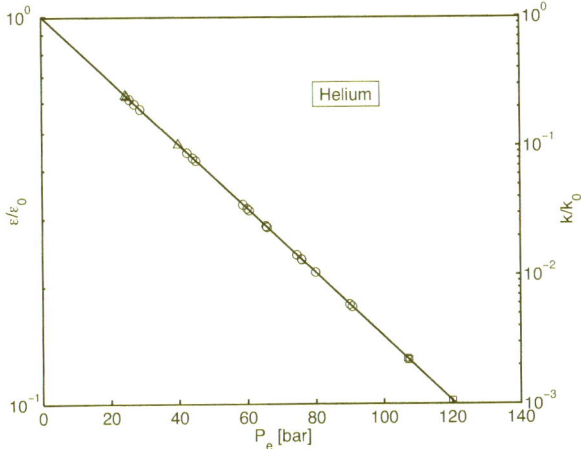

Figure 5.6: Model predicted coal sample relative porosity and permeability as a function of the effective pressure $P_e$ for Helium at 45°C with confining pressure $P_c$ kept constant at (△) 60 , (○) 100 and (□) 140 bar, respectively. Lines are model results and symbols represent the corresponding permeability and porosity obtained at the end of each transient step.

Table 5.4: Porosity and permeability data at 45°C obtained at the end of each transient step when Helium is injected.

| $P_c$ [bar] | $P_{eq}$ [bar] | $\varepsilon$ [%] | $k$ [x $10^3$ mD] |
|---|---|---|---|
| 60 | 20.0 | 1.48 | 5.01 |
|  | 35.1 | 1.97 | 11.82 |
|  | 35.5 | 1.98 | 12.07 |
| 100 | 9.3 | 0.56 | 0.28 |
|  | 10.0 | 0.57 | 0.29 |
|  | 20.1 | 0.69 | 0.51 |
|  | 24.1 | 0.75 | 0.65 |
|  | 25.6 | 0.77 | 0.70 |
|  | 34.2 | 0.90 | 1.15 |
|  | 34.6 | 0.91 | 1.17 |
|  | 39.4 | 1.00 | 1.55 |
|  | 40.0 | 1.01 | 1.60 |
|  | 41.3 | 1.03 | 1.72 |
|  | 54.9 | 1.34 | 3.74 |
|  | 55.8 | 1.36 | 3.93 |
|  | 57.5 | 1.41 | 4.32 |
|  | 71.1 | 1.83 | 9.42 |
|  | 72.8 | 1.88 | 10.36 |
|  | 74.2 | 1.94 | 11.24 |
| 140 | 20.0 | 0.32 | 0.05 |
|  | 32.7 | 0.41 | 0.11 |
|  | 33.1 | 0.41 | 0.11 |

## 5.4 Results and Discussion

an adsorbing fluid allow us to investigate the effects of adsorption and swelling on the flow dynamics. Figure 5.7 reports two examples of transient steps obtained with $CO_2$ when (a) the confining pressure was kept constant at 100 bar and (b) when it was changed between 50 and 100 bar, whereas Figure 5.8 shows one example of transient steps performed with $N_2$ under constant confining pressure.

In both cases the range of gas reservoir pressure tested varied between 10 and 80 bar. The symbols are the experimental values whereas the solid lines correspond to the model results, obtained by fitting the mass transfer coefficient, $k_{M,i}$ and the swelling coefficient, $C_{s,i}$ to the experimental data. In both cases a good agreement is achieved between experiments and simulated transient steps; the corresponding fitted model parameters are summarized in Table 5.3. The mass transfer coefficient is a lumped parameter which combines all the kinetic factors related to the gas diffusion in the coal matrix. Its reciprocal value corresponds to the sorption time constant used in other studies (Bromhal et al., 2005; Shi and Durucan, 2005a), whose values are in agreement with those found in this work, i.e. in the order of a few days.

A reasonably good fit could only be achieved by allowing the coefficient $C_{s,i}$ to be species dependent. It is known, that the coal material properties can change upon sorption (Viete and Ranjith, 2006; Wang et al., 2007), and that, in agreement with the suggestion that coals possess also a polymeric nature, weakening and plasticization phenomena occur with addition of $CO_2$ (Van Krevelen, 1981; Larsen, 2004). Of particular interest here is the lowering of the Young's elastic modulus $E_Y$ when gases are absorbed into the coal's structure (Viete and Ranjith, 2007). Instead of allowing for the Young's modulus to be fluid dependent, which in our model is present as a constant in the swelling term of Eq.(5.11), the coefficient $C_{s,i}$ has been let change depending on the species injected.

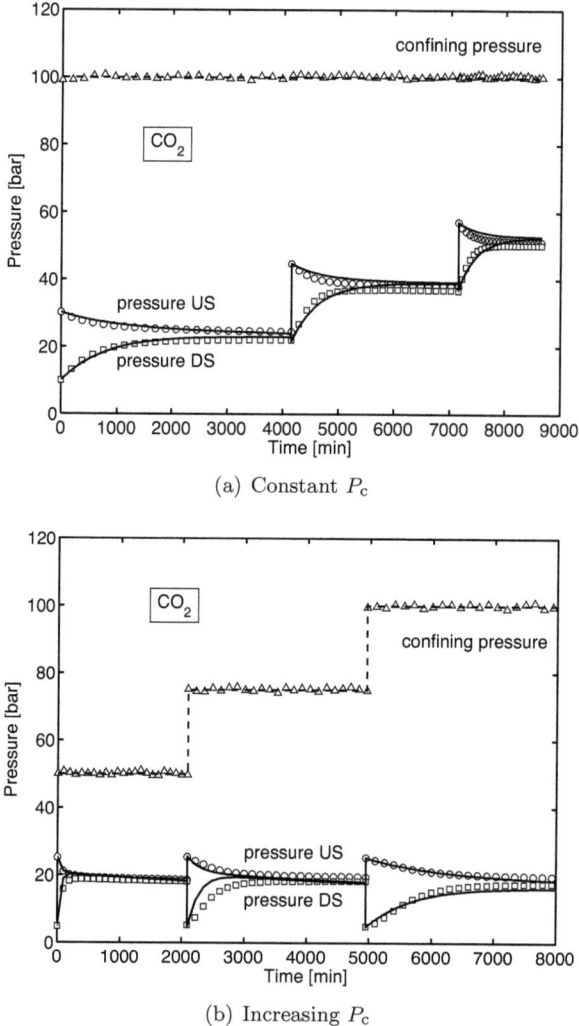

Figure 5.7: Transient steps measured at 45°C when $CO_2$ is injected: (a) confining pressure kept constant and (b) varying confining pressure. Confining pressure $P_c$ ($\triangle$), upstream $P_{US}$ ($\circ$) and downstream $P_{DS}$ ($\square$) reservoir pressures as a function of time. Solid lines correspond to model results.

## 5.4 Results and Discussion

The smaller value found for the coefficient $C_{s,CO_2}$ compared to $C_{s,N_2}$ is therefore compensating for the lowering of $E_Y$ when $CO_2$ is used instead of $N_2$.

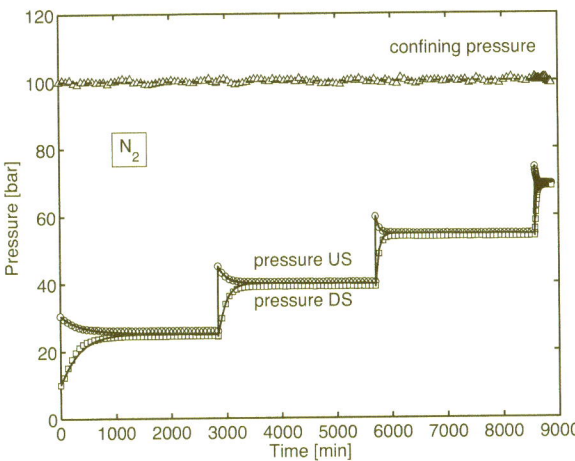

Figure 5.8: Transient steps measured at 45°C when $N_2$ is injected by keeping the confining pressure constant at 100 bar. Confining pressure $P_c$ ($\triangle$), upstream $P_{US}$ ($\circ$) and downstream $P_{DS}$ ($\square$) reservoir pressures as a function of time. Solid lines correspond to model results.

When the two fluids are compared in terms of the same transient step, different equilibration times are measured, with $N_2$ being faster than $CO_2$. Since slower dynamics are a consequence of a lower permeability, this result suggests that the higher degree of sorption and swelling is slowing down the process by closing the cleats. Moreover, as it has been observed with Helium, also with $CO_2$ and $N_2$ the transient step's equilibration time decreases with decreasing effective pressure on the sample. This implies that in Eqs.(5.7) and (5.11) the effective pressure term prevails on the swelling one, resulting in a net increase in permeability. This effect is shown in Figure 5.9, where the obtained relative

5. Permeability of coal

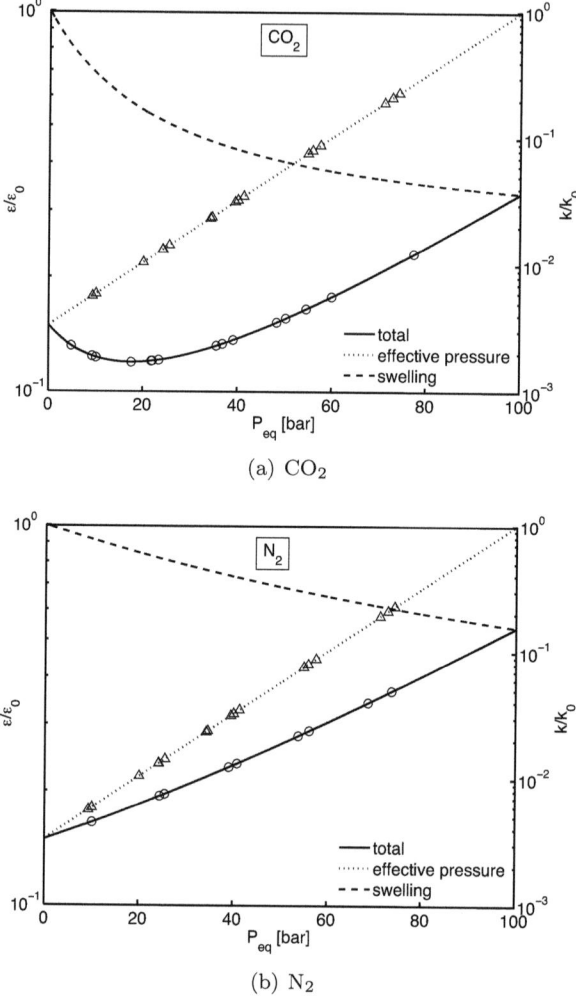

(a) $CO_2$

(b) $N_2$

Figure 5.9: Model predicted relative porosity and permeability as a function of the equilibrium gas pressure $P_{eq}$ of (a) $CO_2$ and (b) $N_2$ with confining pressure kept constant at 100 bar when only the effective pressure term (dotted line), the swelling contribution (dashed line) and both terms (solid line) are taken into account. Lines are model results; symbols are the corresponding permeability and porosity obtained at the end of each transient step: Helium (△) and $CO_2$ or $N_2$ (○).

## 5.4 Results and Discussion

porosity ($\varepsilon/\varepsilon_0$) and permeability ($k/k_0$) curves are reported as function of the equilibrium gas pressure $P_{eq}$ for $CO_2$ and $N_2$, respectively. In both cases, the confining pressure $P_c$ was kept constant at a value of 100 bar. In each figure, three curves are shown corresponding to the cases where in Eq.(5.11) only the terms accounting for effective pressure (dotted line), swelling (dashed line) or both of them are used (solid line). Note that in the first case (dotted line) the obtained relative porosity and permeability curves are identical to the ones for the Helium case at the same confining pressure conditions. Figure 5.9 shows also the permeability and porosity data obtained at the end of each transient step, i.e. when the pressure in the downstream reservoir reaches $P_{eq}$. For the sake of completeness, these data are reported together with the corresponding confining and equilibrium downstream pressures in Table 5.5 and 5.6 in the case of $CO_2$ and $N_2$, respectively. These curves can be used to predict the porosity and permeability of the sample at different equilibrium conditions of gas pressure.

It can be observed that in the case of $CO_2$ a permeability minimum is obtained, which is positioned at the so-called rebound pressure $P_{rb}$ (Palmer and Mansoori, 1998; Shi and Durucan, 2004a). After inserting the pressure dependent Langmuir equation of the swelling $s$ into Eq.(5.11), the rebound pressure can be readily found:

$$P_{rb} = \sqrt{\frac{C_{s,i} E_Y s_{m,i}}{C_e b_i^s} - \frac{1}{b_i^s}} \tag{5.14}$$

and for $CO_2$ it takes a value of about 17 bar. It is worth noting that this minimum is positioned at a pressure which lies in the range of gas pressures tested in this study. At pressure levels below $P_{rb}$, the swelling is strong enough to lower the permeability, whereas at pressure values larger than $P_{rb}$, the pressure effect prevails over the swelling one, and

Table 5.5: Porosity and permeability data at 45°C obtained at the end of each transient step when $CO_2$ is injected.

| $P_c$ [bar] | $P_{eq}$ [bar] | $\varepsilon$ [%] | $k$ [x $10^3$ mD] |
|---|---|---|---|
| 50  | 4.8  | 1.08 | 1.95 |
|     | 18.3 | 0.98 | 1.45 |
| 75  | 5.3  | 0.67 | 0.46 |
|     | 18.3 | 0.61 | 0.35 |
| 100 | 4.9  | 0.42 | 0.11 |
|     | 9.3  | 0.39 | 0.09 |
|     | 10.1 | 0.39 | 0.09 |
|     | 17.6 | 0.38 | 0.08 |
|     | 21.9 | 0.38 | 0.09 |
|     | 22.0 | 0.38 | 0.09 |
|     | 23.4 | 0.38 | 0.09 |
|     | 35.6 | 0.42 | 0.11 |
|     | 36.9 | 0.42 | 0.12 |
|     | 39.2 | 0.43 | 0.13 |
|     | 48.3 | 0.48 | 0.17 |
|     | 50.3 | 0.49 | 0.19 |
|     | 54.6 | 0.52 | 0.22 |
|     | 60.0 | 0.56 | 0.28 |
|     | 77.5 | 0.73 | 0.60 |

Table 5.6: Porosity and permeability data at 45°C obtained at the end of each transient step when $N_2$ is injected.

| $P_c$ [bar] | $P_{eq}$ [bar] | $\varepsilon$ [%] | $k$ [x $10^3$ mD] |
|---|---|---|---|
| 100 | 10.1 | 0.52 | 0.22 |
|     | 24.4 | 0.61 | 0.36 |
|     | 25.5 | 0.62 | 0.37 |
|     | 39.0 | 0.73 | 0.61 |
|     | 40.8 | 0.75 | 0.65 |
|     | 53.8 | 0.88 | 1.07 |
|     | 56.1 | 0.91 | 1.18 |
|     | 68.6 | 1.08 | 1.96 |
|     | 73.6 | 1.16 | 2.42 |

## 5.4 Results and Discussion

therefore the permeability increases. From an ECBM perspective Figure 5.9 shows that the higher the injection pressure the higher the permeability. In the case of $CO_2$ this applies beyond the rebound pressure, in the case of $N_2$ always. However, there is an upper bound for pressure, due on the one hand to the energy penalty associated to pressurization of the feed gas and on the other hand to the requirement of mechanical stability of the coal seam. In practice, injection pressures equal or slightly larger than the prevailing hydrostatic pressure are applied.

### 5.4.3 Experiments with closed system

All the experiments described in the previous sections have been carried out by keeping the confining pressure constant throughout the experiment. In the case of coal swelling upon gas adsorption, this means that the sample is allowed to expand to a certain extent by letting some confining oil in our experimental setup to drain out. On the contrary, in a closed system coal swelling reduces the volume available for the confining fluid and as a consequence the confining pressure increases. The first experimental mode allows defining more precisely the experimental conditions of the experiment, whereas the latter corresponds to a situation which is closer to the more realistic scenario of a zero lateral strain case. Here all the volumetric changes are accommodated by the porosity, with a consequent stronger reduction in the permeability, compared to that observed under constant hydrostatic confinement. Experiments have been carried out to reproduce this situation: after placing the sample into the hydrostatic cell, a confining pressure is applied and the valve connecting the unit that controls the oil pressure is closed: the hydrostatic cell is therefore isolated and all the volumetric changes of the coal sample will therefore be translated into a change of oil pressure (with the assumption that at these pressure levels the oil is incompressible). Transient step experiments have been performed with He and $CO_2$ following the usual procedure and are shown in Figure 5.10.

It can be seen that when Helium is injected, the confining pressure does not change and the situation is therefore identical to the case when the confining pressure is kept constant. On the contrary, when $CO_2$ is injected the confining pressure increases with time, especially at higher gas pressure, i.e. in the second and third transient step. The different behavior observed between the Helium and $CO_2$ experiments can be again interpreted as a consequence of the swelling of the coal core when exposed to an adsorbing gas such as $CO_2$: the oil pressure increase is a response of the increased stress on the oil caused by the swelling of the coal. Moreover, the increase of the confining pressure is greater at larger pressure, like the swelling does with increasing gas pressure. Together with the experimental data, also the model results are shown in Figure 5.10 and in both cases a good agreement is achieved between experiments and simulated transient steps. The measured oil pressures have been interpolated with a polynomial and have been used as input for the confining pressure $P_c$ in the simulation. It is worth pointing out that the model is fully predictive, since the parameters obtained form the experiments with constant oil pressure have been used for the calculations.

## 5.5 Concluding remarks

The study of the aspects related to the fluid flow through coal has a specific practical aspect, namely the assessment of coal seam behavior during an ECBM operation. Laboratory measurements allow in fact to reproduce, or at least to approach, the conditions present in the coal

## 5.5 Concluding remarks

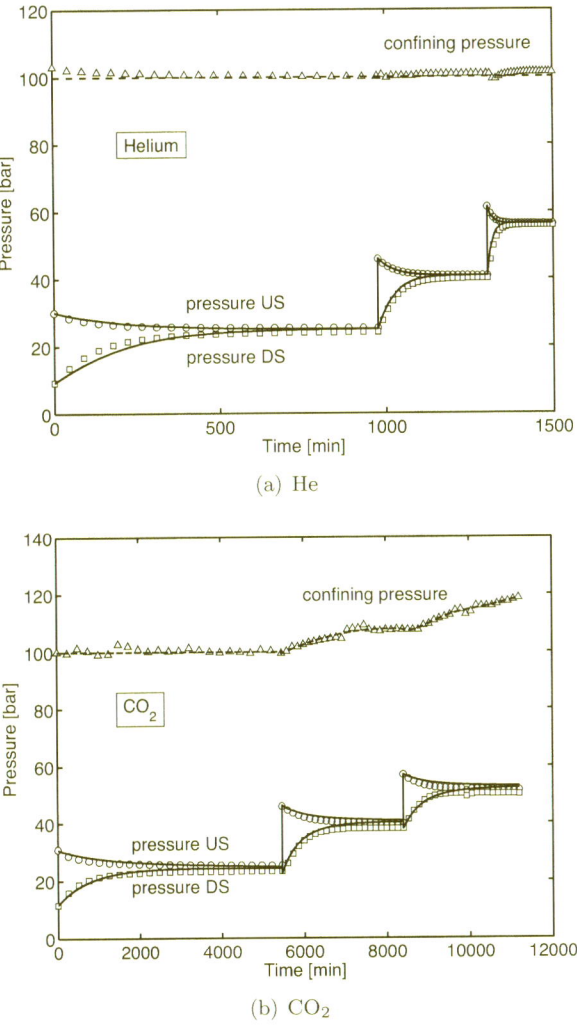

Figure 5.10: Transient steps measured at 45°C when (a) He and (b) $CO_2$ are injected in the closed hydrostatic cell, i.e. without controlling the confining pressure. Confining pressure $P_c$ ($\triangle$), upstream $P_{US}$ ($\circ$) and downstream $P_{DS}$ ($\square$) reservoir pressures as a function of time. Solid lines correspond to model results.

seam. In this study, a technique has been presented which can be used to perform flow experiments on coal cores confined under an external hydrostatic pressure. Moreover, a model describing the fluid flow trough the coal core has been derived, which includes mass balances accounting for gas flow, adsorption and swelling, and mechanical constitutive equations for the description of porosity and permeability changes during injection. The combination of the experimental data with the model predictions allowed to highlight and to understand the different fundamental aspects of the process dynamics related to effective pressure, adsorption and swelling phenomena. In particular, it has been observed that the permeability decreases at increasing effective stress on the sample, and that this effect is even larger when swelling of the coal sample due to gas adsorption occurs. The presented approach provides a way to estimate parameter values which can be directly used in 3D models typically used to simulate reservoir dynamics.

However, there are some important issues that need to be considered before extending the results obtained in the laboratory to a real case. The size of the sample used in the present study is extremely small compared to the dimension of a coal seam; even if an extensive sampling is carried out, such dimensions preclude the coverage of the complete coal fracture system, in particular of the larger fractures (Wang et al., 2007). As a consequence, the obtained flow rates and permeability are underestimated with respect to those observed in a coal seam. Moreover, the experiments have been carried out under a constant hydrostatic confining pressure, which doesn't represent exactly the underground *in-situ* condition: it is known that the stresses in a coal seam have a different magnitude for each orthogonal direction, suggesting that the changes mentioned above do possess indeed a three dimensional nature (Gray, 1987). Experiments with a triaxial Heard type rig should therefore be

## 5.5 Concluding remarks

preferred and will be the focus of next studies. Finally, an aspect to be considered is the presence of water in the coal seam. Beside reducing the adsorption of $CO_2$ and therefore the total storage capacity of the coal seam, it complicates considerably the description of the problem and displacement dynamics.

Therefore, the conclusions reached in this study can be summarized as follows. Laboratory studies coupled with a physically sound modeling work are very effective to understand the different mechanisms acting during an ECBM operation. In addition, future studies should address the open issues mentioned above, thus allowing to better reproduce field conditions in the laboratory. However, lab scale studies can be not enough to predict the behavior of a real system and the laboratory research has to be carried out in parallel to field studies. This will allow to extend the lab and the modeling work to a larger scale for a more reliable description of the real reservoir, therefore increasing the confidence on ECBM for its deployment at a commercial scale.

## 5.6 Nomenclature

| | |
|---|---|
| $A$ | Cross sectional area [m$^2$] |
| $b$ | Langmuir constant (adsorption isotherm, $q(c)$) [m$^3$/mol] |
| $b^\mathrm{p}$ | Langmuir constant (adsorption isotherm, $q(P)$) [Pa$^{-1}$] |
| $b^\mathrm{s}$ | Langmuir constant (swelling isotherm, $s(P)$) [Pa$^{-1}$] |
| $c$ | Fluid phase concentration [mol/m$^3$] |
| $C_\mathrm{c}$ | Confining pressure coefficient [-] |
| $C_\mathrm{e}$ | Effective pressure coefficient [-] |
| $C_\mathrm{p}$ | Pore pressure coefficient [-] |
| $C_\mathrm{o}$ | Coefficient boundary conditions [Pa] |
| $e$ | Bulk volumetric strain [-] |
| $E_{jk}$ | Strain tensor [Pa] |
| $E_\mathrm{Y}$ | Young's elastic modulus [Pa] |
| $g$ | Gravitational acceleration [m/s$^2$] |
| $k$ | Permeability [mD] (1 D = 9.869233×10$^{-13}$ m$^2$) |
| $k_\mathrm{M}$ | Mass transfer coefficient [s$^{-1}$] |
| $K$ | Bulk modulus [Pa] |
| $L$ | Coal core length [cm] |
| $M_\mathrm{m}$ | Molar mass of adsorbate [g/mol] |
| $N_\mathrm{exp}$ | Number of experimental data points [-] |
| $P$ | Pressure [bar] |
| $q$ | Solid phase concentration [mol/m$^3$] |
| $q^*$ | Equilibrium solid phase concentration [mol/m$^3$] |
| $q_\mathrm{m}$ | Saturation capacity (adsorption isotherm, $q(c)$) [mol/g] |
| $q_\mathrm{m}^\mathrm{p}$ | Saturation capacity (adsorption isotherm, $q(P)$) [mol/g] |
| $R_\mathrm{o}$ | Vitrinite reflectance coefficient [-] |
| $s$ | Swelling [-] |
| $s_\mathrm{m}$ | Saturation capacity (swelling isotherm, $s(P)$) [-] |

## 5.6 Nomenclature

| | |
|---|---|
| $t$ | Time [s] |
| $T_{jk}$ | Stress tensor [Pa] |
| $T$ | Mean normal stress [Pa] |
| $u$ | Superficial velocity [m/s] |
| $V$ | Volume [cm$^3$] |
| $z$ | Axial coordinate [m] |

**Greek Letters**

| | |
|---|---|
| $\gamma$ | Average rock mass density [Pa/m] |
| $\delta_{jk}$ | Dirac's delta [-] |
| $\varepsilon$ | Porosity [-] |
| $\mu$ | Fluid dynamic viscosity [Pa·s] |
| $\nu$ | Poisson's ratio [-] |
| $\rho_{\text{ads}}$ | Adsorbent (coal) density [g/cm$^3$] |
| $\tau_{0.5}$ | Time to reach $P_{0.5}$ [s] |

**Subscripts and Superscripts**

| | |
|---|---|
| c | Confining |
| e | Effective |
| eq | Equilibrium |
| exp | Experiment |
| DS | Downstream |
| $i$ | Component $i$ |
| $j, k$ | Axial coordinates $x, y, z$ |
| mod | Model |
| US | Downstream |

| | |
|---|---|
| p | Gas pressure |
| rb | Rebound |
| s | Swelling |
| 0 | Initial (unstressed) state |

# Chapter 6

# Coal bed dynamics

## 6.1 Introduction

As seen in the previous chapter, gas sorption and swelling have complex effects on the variation of coal porosity and permeability and therefore on the performance of an ECBM operation. An accurate description of the $CO_2/CH_4$ displacement dynamics in the coal seam is essential for the development of reliable reservoir simulators used to history match field test data obtained from ECBM field tests. Input for these models are the results of laboratory studies, which have focused on the different aspects related to $CO_2$ storage in coal seams, namely gas sorption, coal swelling and the issue related to the permeability reduction upon gas injection. These elements, which have been treated separately in the previous chapters of this thesis, are strongly dependent on each other; the objective of this and other modeling studies is to integrate them into a suitable mathematical model in order to investigate how their simultaneous acting is affecting the overall picture of the ECBM process. Outcomes of these modeling studies are estimates of several parameters, such as the amount of $CO_2$ which can be stored in the coal bed, the amount and the purity of $CH_4$ recovered, and the time needed for the $CO_2$ to break through at the production well. Any ECBM operation's design is based on the values of these parameters and therefore, once validated, these models can be used as a tool to critically assess the success or the failure of the performed field tests and to plan future demonstration projects.

One dimensional models have been shown to provide a very useful understanding of the key mechanisms that affect the storage and recovery process (Gilman and Beckie, 2000; Shi and Durucan, 2003; Seto, 2007; Wang et al., 2007; Jessen et al., 2008). Coal reservoirs are fractured systems consisting of a low permeability matrix and a high permeability fracture network. One can distinguish up to four types of pores in coal, namely cleats where gas and water are present, macro- and mesopores where there is only free gas, and micropores where adsorption takes place. The complexity of this pore structure impacts also mass transfer mechanisms and how to describe them in ECBM models. The general assumption is that the displacement of $CH_4$ by $CO_2$ results from a multistep process. The gas injected in the coal bed diffuses from the fracture network, through the matrix and macropores and finally to the internal surface of the coal. Here, partial pressure with respect to the adsorbed gas is reduced, causing desorption, and gas exchange takes place. The desorbed gas diffuses through the matrix and micropores, out to the fracture network where it flows to the production well (Gentzis, 2000; Totsis et al., 2004; Seto, 2007). Traditionally, this mass transfer is described through a linear driving force model either by lumping gas diffusion in the different types of pore using a single mass transfer coefficient,

## 6.1 Introduction

or the corresponding time constant (Bromhal et al., 2005; Sams et al., 2005), or by using a bidisperse pore diffusion model and the corresponding time constants for diffusion in macro-/mesopores and in micropores (Shi and Durucan, 2003, 2005a). Recently, a more detailed model has been proposed, which accounts for the coal pore size distribution and for three mechanisms of coal swelling (Wang et al., 2007). Mass transfer includes convective flow in cleats, convective and diffusive flows in meso- and macropores, adsorption and surface diffusion in micropores, whereby diffusion is described using the Maxwell-Stefan equations. A more compact version of such a model has been shown to describe with good agreement a number of displacement experiments carried out on coal cores in different labs (Wei et al., 2007a,b).

In the previous chapter, a model was developed and successfully applied for describing pure gas injection experiments into coal cores confined under an external hydrostatic pressure and under simulated reservoir temperature and pressure conditions (Pini et al., 2009). The model consists of mass balances accounting for gas flow and sorption, and a constitutive stress-strain relationship for the description of porosity and permeability changes during injection. In the present study, this model is extended to the multicomponent single-phase (gas) displacement in a coal seam. Previous studies have shown that injection of flue gas into a coal bed for ECBM recovery is a promising alternative to pure $CO_2$ injection for several reasons (Bustin et al., 2008; Durucan and Shi, 2009). First, being flue gas the combustion exhaust gas produced by power plants, it could be directly injected, thus avoiding the expensive capture step. Secondly, since flue gas consists of mostly nitrogen (87%) and carbon dioxide (13%), the presence of the former allows keeping the coal permeability sufficiently high.

In this study, numerical simulations on the performance of $CO_2$ storage and ECBM recovery in coal beds are presented. In particular, different ECBM scenarios involving the injection of gas mixtures with different composition (from pure $N_2$ to pure $CO_2$) into a coal bed previously saturated with methane are investigated. The performance of each scenario is then compared in terms of amount of $CO_2$ stored as well as amount and purity of $CH_4$ produced.

## 6.2 Modeling

The one-dimensional core model presented in the previous chapter and used to describe the pure gas injection experiments is now extended to mixtures. This will allow us to study the gas flow process in coal beds and the displacement dynamics during an ECBM operation. During primary recovery, the coal bed methane is recovered by reducing the hydrostatic pressure of the seam through dewatering. After an initial stage where mainly water is produced, the flow regime becomes bi-phasic (gas/water) with the amount of produced water declining in time (Durucan and Shi, 2009). In the final stage of primary recovery the amount of produced water becomes practically insignificant and any water present the reservoir can be considered immobile (Zhu et al., 2003). We consider the situation where ECBM production starts at the end of the primary recovery operation, where the coal bed behave almost as a dry reservoir, and therefore single-phase (gas) flow can be assumed.

Coal reservoirs are fractured systems consisting of a low permeability matrix and a high permeability fracture network. As in the case of the model presented in Chapter 5, the coal total porosity $\varepsilon^*$ is divided into cleat porosity $\varepsilon$ and macroporosity $\varepsilon_p$, with the microporosity being accounted for as combined with the solid material, i.e.

## 6.2 Modeling

$$\varepsilon^* = \varepsilon + (1-\varepsilon)\varepsilon_\text{p} \tag{6.1}$$

Mass transfer is then described through a linear driving force model by lumping gas diffusion in the different types of pores using a single mass transfer coefficient or the corresponding time constant (Bromhal et al., 2005; Sams et al., 2005).

### 6.2.1 Mass balances

With reference to Chapter 5, material balances can be written for each component $i$ in the fluid and in the adsorbed phase:

$$\frac{\partial(\varepsilon^* c_i)}{\partial t} + \frac{\partial[(1-\varepsilon^*)q_i]}{\partial t} + \frac{\partial(uc_i)}{\partial z} = 0 \tag{6.2a}$$

$$\frac{\partial[(1-\varepsilon^*)q_i]}{\partial t} = (1-\varepsilon^*)k_{\text{M},i}(q_i^* - q_i) \tag{6.2b}$$

where $c_i$ and $q_i$ are respectively the gas and adsorbed phase concentration of component $i$; $q_i^*$ is its equilibrium concentration in the adsorbed phase, $k_{\text{M},i}$ is its mass transfer coefficient, $u$ is the superficial velocity, and $t$ and $z$ are time and space coordinates. The superficial velocity $u$ is given by Darcy's law, which expresses velocity as a function of pressure gradient and permeability:

$$u = v\varepsilon = -\frac{k}{\mu}\left(\frac{\partial P}{\partial z}\right) \tag{6.3}$$

where $v$ is the interstitial velocity, $P$ the total pore pressure, $k$ the permeability and $\mu$ the dynamic viscosity. As explained in the next section, the porosity, and as a consequence the permeability, are changing depending on the specific stress situation in the coal bed, which is affected by the fluid pressure. The sorption on coal is described by an extended Langmuir equation, giving the amount adsorbed for component $i$, $q_i^*$, as a function of its concentration:

$$q_i^* = \frac{\rho_{\text{ads}} q_{m,i} b_i y_i P}{1 + P \sum_{j=1}^{N_c} b_j y_j} \qquad (6.4)$$

where $y_i$ the molar fraction, $N_c$ the number of components and $\rho_{\text{ads}}$ the coal bulk density; $q_{m,i}$ and $b_i$ are the saturation capacity per unit volume adsorbent and the Langmuir equilibrium constant, respectively. In Figure 6.1 are shown the $CO_2$, $CH_4$ and $N_2$ Langmuir sorption isotherms as a function of the pressure $P$ for the Italian coal from the Sulcis Coal Province, whose sorption data have been presented in Chapter 3. Values for the fitted parameters are reported in Table 6.1.

The model is completed by the following constitutive equations: the Peng-Robinson EOS, needed to relate gas density to pressure and temperature (Peng and Robinson, 1976) and whose parameters are given in Table 6.2, and a relationship for the gas mixture viscosity following the method of Wilke (Reid et al., 1987). Finally, initial and boundary conditions are defined as follows:

$$\text{Initial conditions}: \quad \text{at } t = 0 \quad \begin{array}{ll} c_i = c_i^0 & 0 \leq z \leq L \\ q_i = q_i^0 & 0 \leq z \leq L \end{array}$$

## 6.2 Modeling

$$\text{Boundary conditions:} \quad \text{at } z = 0 \quad c_i = c_i^{\text{inj}} \quad t > 0$$
$$\text{at } z = L \quad P = P_{\text{out}} \quad t > 0$$

Table 6.1: Langmuir constants for the sorption and swelling isotherms for the coal from the Sulcis Coal Province.

|        | Sorption isotherm     |                        | Swelling isotherm     |                        |
|--------|-----------------------|------------------------|-----------------------|------------------------|
|        | $q_{m,i}$ [mol/g]     | $b_i$ [Pa$^{-1}$]      | $s_{m,i}$ [-]         | $b_i^s$ [Pa$^{-1}$]    |
| $CO_2$ | $2.49 \times 10^{-3}$ | $1.25 \times 10^{-6}$  | $4.90 \times 10^{-2}$ | $3.80 \times 10^{-7}$  |
| $CH_4$ | $1.56 \times 10^{-3}$ | $6.26 \times 10^{-7}$  | $2.33 \times 10^{-2}$ | $3.47 \times 10^{-7}$  |
| $N_2$  | $1.52 \times 10^{-3}$ | $1.40 \times 10^{-7}$  | $1.70 \times 10^{-2}$ | $5.19 \times 10^{-8}$  |

Table 6.2: Thermodynamic properties of $CO_2$, $CH_4$ and $N_2$ for the Peng-Robinson EOS.

|        | $T_c$   | $P_c$   | $w$     | $k_{ij}$ |        |        |
|--------|---------|---------|---------|----------|--------|--------|
| Fluid  | [K]     | [bar]   | [-]     | $N_2$    | $CH_4$ | $CO_2$ |
| $N_2$  | 126.192 | 33.958  | 0.0372  | 0        | 0.031  | -0.02  |
| $CH_4$ | 190.56  | 45.992  | 0.01142 | 0.031    | 0      | 0.103  |
| $CO_2$ | 304.13  | 73.773  | 0.22394 | -0.02    | 0.103  | 0      |

### 6.2.2 Stress-strain relationship

A stress-strain constitutive equation is required to describe the mechanical behavior of the coal bed under reservoir conditions. The fluid pressure in the coal bed plays a decisive role in determining the stress situation of the reservoir, thus affecting markedly the permeability of the porous medium (Gray, 1987; Cui et al., 2007). First, fractures are widened or closed, depending on whether the pressure is increased or reduced. Secondly, upon gas sorption the coal swells thus reducing the fracture openings. It was shown in Chapter 4, that coal swelling can be effectively described by a Langmuir-like equation, in agreement with

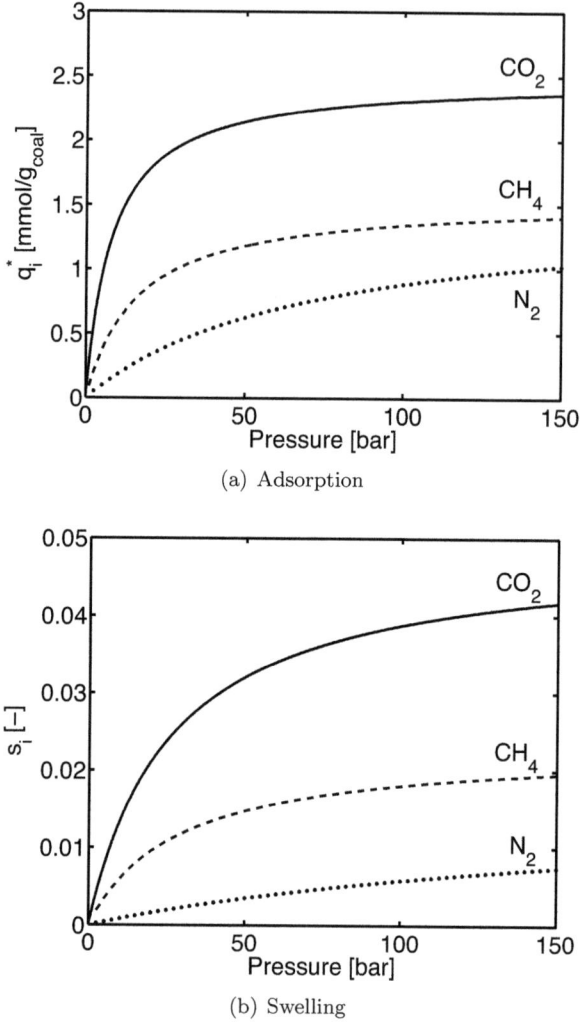

Figure 6.1: Langmuir sorption (a) and swelling (b) isotherms at 45°C as a function of pressure for $CO_2$ (solid line), $CH_4$ (dashed line) and $N_2$ (dotted line) for a coal from the Sulcis coal province.

## 6.2 Modeling

other studies (Levine, 1996; Palmer and Mansoori, 1998; Shi and Durucan, 2004a; Cui et al., 2007; Pini et al., 2009). In an analogous way as for sorption of gas mixtures, an extended Langmuir equation for swelling can be expressed as follows:

$$s_i = \frac{s_{m,i} b_i^s y_i P}{1 + P \sum_{j=1}^{N_c} b_j^s y_j} \quad (6.5)$$

Figure 6.1 shows the $CO_2$, $CH_4$ and $N_2$ Langmuir swelling isotherms as a function of the pressure $P$ for the Italian coal from the Sulcis Coal Province, whose swelling data have been presented in Chapter 4. Values for the fitted parameters are reported in Table 6.1. The swelling equation, Eq.(6.5), can be combined with the sorption isotherm, Eq.(6.4), to obtain the swelling as a function of the adsorbed amount instead of pressure. This has the advantage of accounting for the kinetic of the swelling process through the adsorption rate given by Eq.(6.2b), as explained in Pini et al. (2009). In its final form, the total swelling is defined by the following equation,

$$s = \frac{\sum_{i=1}^{N_c} C_{s,i} \beta_i \alpha_i q_i}{C_s \left(1 - \sum_{j=1}^{N_c} \alpha_i q_i\right)} \quad (6.6)$$

where the parameters $\alpha_i$ and $\beta_i$ are functions of the Langmuir parameters of the sorption and swelling isotherms, i.e.

$$\alpha_i = \frac{b_i - b_i^s}{q_{m,i} b_i} \quad (6.7a)$$

$$\beta_i = \frac{b_i^s s_{m,i}}{b_i - b_i^s} \qquad (6.7b)$$

Note that Eq.(6.6) is valid for $0 \leq q_i \leq q_{m,i}$.

The behavior of the coal bed permeability during gas injection is taken into account through implementation of a dynamic permeability model. In agreement with other models reported in the literature (Gilman and Beckie, 2000; Shi and Durucan, 2004a; Bustin et al., 2008), the permeability relationship validated in the previous chapter takes the following general form:

$$\frac{k}{k_0} = \left(\frac{\varepsilon}{\varepsilon_0}\right)^3 = \exp\left[-C_1(P_c - P) - C_2 s\right] \qquad (6.8)$$

where $P$ and $P_c$ are fluid and confining pressure, respectively; $s$ is the pressure dependent total swelling; $C_1$ and $C_2$ are coefficients depending on coal properties. Two terms can be recognized in Eq.(6.8): one depends on the effective pressure exerted on the coal (defined as the confining pressure minus the fluid pressure) and the other on the swelling/shrinkage of the coal upon gas sorption. The resulting net effect determines whether the permeability would be enhanced or reduced compared to an arbitrarily chosen initial state (defined with the subscript 0). In the above equation, the reference values of porosity and permeability apply to an unstressed coal in contact with a non-swelling gas at atmospheric pressure (Pini et al., 2009). For the sake of better visualization, Figure 6.2 shows the permeability ratio as a function of the $CO_2$ fluid pressure for a constant confining pressure of 100 bar. The three curves correspond to the case where in Eq.(6.8) only the term accounting for the effective pressure (dotted line), swelling (dashed line) or both of them are used (solid line). It can be seen that the permeability increases with fluid pressure, due to a decrease in the effective

## 6.2 Modeling

stress (dotted line). As expected, as the fluid pressure reaches the same value as the confining pressure, the permeability is completely recovered, i.e. $k = k_0$. However, the swelling of the coal upon $CO_2$ sorption closes the coal cleats, thus counteracting this permeability enhancement. The combination of both effects leads to a characteristic behavior of the permeability curve, where, at the so-called rebound pressure, the permeability reaches its minimum value, and after which it increases by further increasing the pressure (Palmer and Mansoori, 1998; Shi and Durucan, 2004a). As explained in Chapter 5, the extent to which the coal swells when exposed to a given fluid determine whether the minimum appears or not in the range of pressures investigated.

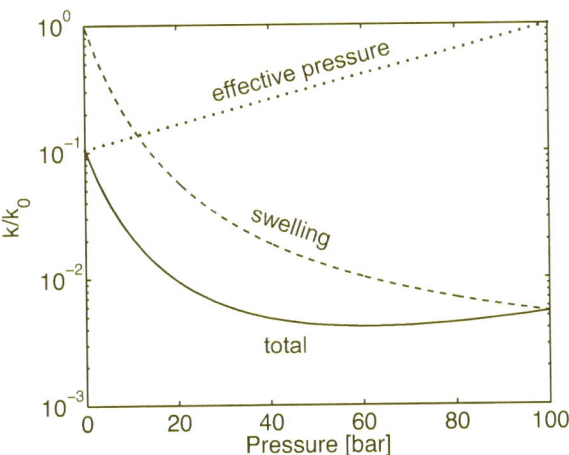

Figure 6.2: Model predicted relative permeability as a function of the $CO_2$ gas pressure $P$ with confining pressure kept constant at 100 bar when only the effective pressure term (dotted line), the swelling contribution (dashed line) and both terms (solid line) are taken into account ($C_1$=225.7 GPa$^{-1}$ and $C_2$=134.4).

The parameters $C_1$ and $C_2$ in the permeability Eq.(6.8) can be estimated based on mechanical properties only, under the assumption of a specific simplified stress situation of the coal bed. In particular, two situations have been investigated in the literature, namely a condition of uniaxial strain and constant overburden stress (Bustin et al., 2008) and the same situation under the assumption that deformation takes place only in the horizontal direction (Gilman and Beckie, 2000; Shi and Durucan, 2004a). Another method to estimate these parameters is by fitting them to experimental permeability data, as explained in Chapter 5. In Table 6.3 are reported the relationships for the constants $C_1$ and $C_2$ for these three situations. With the known definition of the bulk modulus, i.e $K = E_Y/[3(1-2\nu)]$, it can be seen that for all three models, the input parameters are the two coal elastic properties, namely the Young's modulus $E_Y$ and Poisson's ratio $\nu$, respectively. In the model proposed by Gilman and Beckie (2000), $E_f$ is some analogous of Young's modulus for a fracture, whereas in the model by Shi and Durucan (2004a), $C_f$ is defined as the fracture compressibility. Both parameters can be found by fitting them to experimental data. In an analogous way, in the model by Pini et al. (2009), which is also used in this study, experiments with a non-adsorbing (and non-swelling) gas, were used to obtain values for $C_e$, whereas experiments with an adsorbing gas, such as $CO_2$ were then used to estimate the values for the coefficient $C_s$.

## 6.2.3 Solution procedure

The problem is defined by Eqs.(6.2)-(6.4), (6.6) and (6.8). The orthogonal collocation method has been applied to discretize in space the partial differential equations (Villadsen and Michelsen, 1978; Morbidelli et al., 1983). The resulting system of ordinary differential equations has then

## 6.2 Modeling

Table 6.3: Constants $C_1$ (Pa$^{-1}$) and $C_2$ of Eq.(6.8) as obtained from different permeability models.

| Reference | $C_1$ | $C_2$ |
|---|---|---|
| Gilman and Beckie (2000) | $\frac{3\nu}{E_f(1-\nu)}$ | $\frac{3E_Y}{(1-\nu)E_f}$ |
| Shi and Durucan (2004a) | $\frac{3C_f\nu}{(1-\nu)}$ | $\frac{C_f E_Y}{(1-\nu)}$ |
| Bustin et al. (2008) | $\frac{1+\nu}{K\varepsilon_0(1-\nu)}$ | $\frac{2E_Y}{3(1-\nu)K\varepsilon_0}$ |
| Pini et al. (2009) | $\frac{3C_e}{K\varepsilon_0}$ | $\frac{3C_s E_Y}{K\varepsilon_0}$ |

been solved numerically using a commercial ODEs solver (in Fortran). The input parameters used for the model calculations are summarized in Table 6.4. A situation representative for a coal bed lying at 500 m depth is described, whose properties are those of the Italian Coal of the Sulcis Coal Province (Sardinia, Italy). The coal bed permeability has been chosen to match typical values for coal seams, which lie between 1 and 10 mD (Gilman and Beckie, 2000; White et al., 2005). For all the species a mass transfer coefficient of $10^{-5}$ s$^{-1}$ has been chosen, corresponding to a sorption time constant of about 1.2 days, in agreement with values typically used in reservoir simulators (Bromhal et al., 2005; Shi and Durucan, 2005a; Shi et al., 2008). The injection pressure ($P_{inj}$=40 bar) is chosen to by slightly lower than the hydrostatic pressure corresponding to the coal seam depth (50 bar) and the pressure at the production well is kept constant at a value of $P_{out}$=1 bar. Moreover, at the beginning of the injection, the reservoir pressure (100% CH$_4$) is lower than the hydrostatic pressure and takes a value of $P_0$=15 bar, as might occur after the coal bed primary production.

Two cases have been investigated, which differ in the value of the parameter $C_2$ in Eq.(6.8), to highlight the effect of the permeability variation on the gas flow dynamics during the ECBM process. For "Case A", the

Table 6.4: Input parameters for the model.

| Property | Value |
|---|---|
| Temperature, $T$ [°C] | 45 |
| Coal seam lenght, $L$ [m] | 100 |
| Coal seam depth [m] | 500 |
| Initial pressure, $P_0$ [bar] | 15 |
| Initial gas composition (% $CH_4$) | 100 |
| Initial permeability, $k_0$ [mD] | 10 |
| Coal bulk density, $\rho_{ads}$ [kg/m$^3$] | 1356.6 |
| Mass transfer coeff, $k_{M,i}$ [s$^{-1}$] | $10^{-5}$ |
| Sorption time, $\tau$ [days] | 1.2 |
| Injection pressure, $P_{inj}$ [bar] | 40 |
| Production pressure, $P_{out}$ [bar] | 1 |

values for the parameter $C_{s,i}$ obtained for each component $i$ from the experiments presented in Pini et al. (2009) have been used to calculate the total swelling in Eq.(6.6) and to obtain the value of the parameter $C_2$. For "Case B", a four times larger value $C_{s,i}$ has been set for $CO_2$ and has been used also for all other components. For this reason, we will refer to this situation as the strong swelling case. The value of these parameters are summarized in Table 6.5. For the sake of comparison, they are reported together with values given in other studies using a similar stress-strain relationship for the permeability. It is worth pointing out that the initial porosity values used in this work are much larger than those from other studies. This is mainly due to the fact that the reference condition (0) is different: in our study it refers to a unstressed state (no confinement, no fluid pressure), whereas in the other studies it refers to the initial reservoir condition, thus taking into account also the overburden stress.

As a measure to compare the outcomes of the different ECBM simulations, the following variables are defined ($i = CH_4$):

## 6.2 Modeling

$$\text{Gas In Place (GIP) [mol/m}^2] = L\left[\varepsilon^* c_i^0 + (1-\varepsilon^*) q_i^0\right] \quad (6.9\text{a})$$

$$\text{Recovered CH}_4 \; [-] = \frac{\int_0^t uc_i \,|_{z=L}\, dt}{\text{GIP}} \quad (6.9\text{b})$$

$$\text{CH}_4 \text{ purity } [-] = \frac{c_i \,|_{z=L}}{\sum_{j=1}^{N_c} c_j \,|_{z=L}} \quad (6.9\text{c})$$

whereas for $CO_2$, the following variables are introduced ($i = CO_2$):

$$\text{Injected CO}_2 \; [\text{mol/m}^2] = \int_0^t uc_i \,|_{z=0}\, dt \quad (6.10\text{a})$$

$$\text{Stored CO}_2 \; [\text{mol/m}^2] = \int_0^L \left(\varepsilon^* c_i + (1-\varepsilon^*) q_i\right) dz \quad (6.10\text{b})$$

Table 6.5: Parameters for the permeability relationship, Eq.(6.8). [a]Ref. Shi and Durucan (2004a, 2006); Shi et al. (2008).

| Parameter | Shi and Durucan[a] | Bustin et al. (2008) | Pini et al. (2009) Case A | Case B |
|---|---|---|---|---|
| $\nu$ [-] | 0.35 | 0.3 | 0.26 | 0.26 |
| $E_Y$ [GPa] | 2.62-2.90 | 3.00 | 1.12 | 1.12 |
| $K$ [GPa] |  | 2.50 | 0.78 | 0.78 |
| $\varepsilon_0$ [-] | 0.001-0.004 | 0.0023 | 0.08 | 0.08 |
| $C_f$ [GPa$^{-1}$] | 116-290 |  |  |  |
| $C_e$ [-] |  |  | 4.676 | 4.676 |
| $C_s$ [-] |  |  | 0.622-2.377 | 2.5 |
| $C_1$ [GPa$^{-1}$] | 187.4-468.5 | 128.1-323.0 | 225.7 | 225.7 |
| $C_2$ [-] | 467.6-1293.9 | 197.0-496.9 | 33.6-128.4 | 134.4 |

## 6.3 Results

### 6.3.1 Permeability changes

By assuming that the methane is completely displaced by the injected gas mixture, the changes in permeability can be analytically estimated with Eq.(6.8) only. Figure 6.3 shows the obtained variations in permeability under different injection schemes (from pure $CO_2$ to pure $N_2$ injection) for Cases A and B. In the figures, the predicted $CO_2/N_2$ curves are compared to the primary recovery scenario (pure $CH_4$, dashed line), for which the coal bed situation before starting gas injection is marked with a circle. The vertical dotted line at 40 bar represents a theoretical abandonment scenario, where, at the end of the ECBM operation, the coal seam has been completely filled with the injected gas at a pressure corresponding to the injection pressure. Qualitatively there is no difference between Case A and B: pure $CO_2$ injection lead to the strongest reduction in permeability, whereas addition of $N_2$ to the mixture allows counteracting this effect. As explained in Section 6.2, the reduction in permeability is controlled by the extent of swelling of the coal, which is fluid dependent. Because of the weak sorption and swelling of $N_2$ compared to $CH_4$ and $CO_2$, the injection of $CO_2/N_2$ mixtures induces less permeability reduction compared to pure $CO_2$. For gas mixtures rich in $N_2$, permeability can be even larger when compared to the initial situation. Note that in Case B the changes in permeability are much larger compared to Case A, as in the former case the y-axis is plotted on a logarithmic scale. In particular, in Case B, for which a larger swelling constant $(C_2)$ in Eq.(6.8) has been used, permeability can either be enhanced or reduced of about one order of magnitude, depending on whether pure $N_2$ or $CO_2$ are injected. Moreover, in Case A the rebound in permeability can be clearly seen, whereas for Case B the

## 6.3 Results

rebound doesn't appear in the pressure range investigated. These results agrees with the observations from the performed field tests, where it was shown that $CO_2$ injection yields to injectivity problems caused by the aforementioned reduction in permeability (Gunter et al., 2004; Reeves, 2004; Van Bergen et al., 2006). It is in fact expected that the main loss in permeability is confined around the injection well, where the $CO_2$ concentration is at the highest. Several attempts have been made to counteract the low injection rates caused by $CO_2$ swelling: at the RECOPOL project a frac-job allowed to substantially increase the injectivity, at least temporarily (Van Bergen et al., 2006). In the Alberta $CO_2$ ECBM project shut-in periods were enforced, with the aim of reducing the gas pressure close to the well (Gunter et al., 2004). Moreover, it was shown that during the injection of flue gas a steady increase of well injectivity was observed (Gunter et al., 2004). This last option is very attractive for several reasons. First, it allows injecting flue gas directly without the expensive $CO_2$ capture step and at the same time keeping the permeability sufficiently high. Secondly, compared to the injection of pure $N_2$, it has the added-value of simultaneously store $CO_2$ in the coal bed.

### 6.3.2 ECBM recovery and $CO_2$ storage

In this section the results of ECBM simulations are presented. In particular, a number of ECBM schemes involving the injection of $CO_2/N_2$ gas mixtures with different composition are compared in terms of performance of the ECBM/$CO_2$ storage operation. Two situations are investigated, which differ in the extent of the swelling term in Eq.(6.8), which leads to a different permeability behavior of the coal bed. Independently of the whether one or the other case is studied, it is important to un-

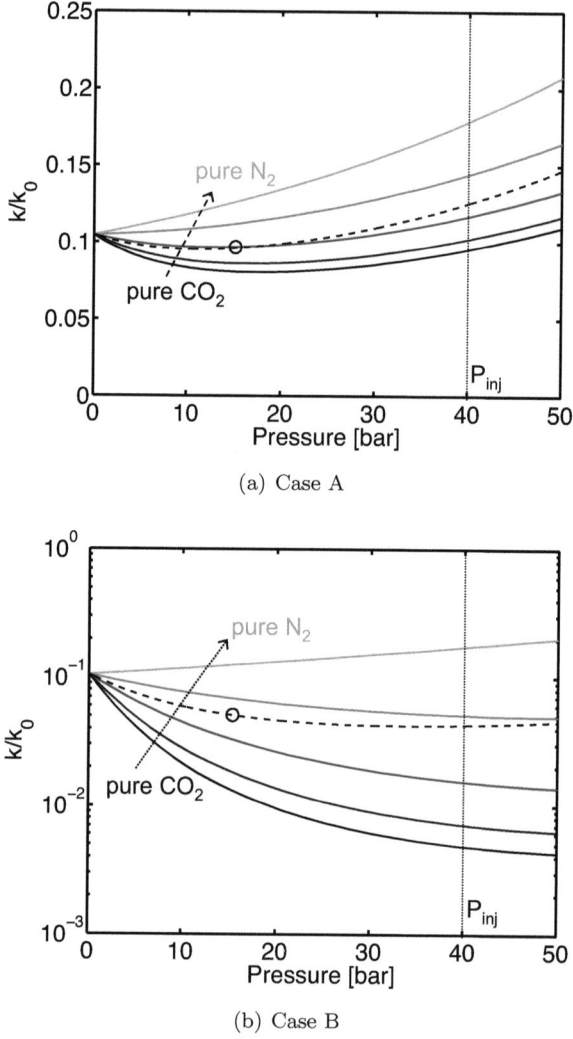

Figure 6.3: Permeability ratio $k/k_0$ as a function of pressure $P$ under different injection scenarios (solid lines, Pure $CO_2$, 80:20/$CO_2$:$N_2$, 50:50/$CO_2$:$N_2$, 20:80/$CO_2$:$N_2$, pure $N_2$) for Case A (a) and Case B (b). The dashed line corresponds to the primary recovery scenario (pure $CH_4$).

## 6.3 Results

derstand first the dynamics of the displacement when pure $CO_2$, pure $N_2$ or a mixture of them is injected. Figure 6.4 shows the concentration profiles of $CO_2$, $CH_4$ and $N_2$ along the coal bed axis for three different injection scenarios: pure $CO_2$ (a) pure $N_2$ (b) and 50:50/$CO_2$:$N_2$ (c). It can be seen that injection of pure $CO_2$ displaces the $CH_4$ through a sharp front, due to the higher adsorptivity of the former compared to the latter. As the preferentially adsorbing $CO_2$ propagates through the coal bed therefore, no $CH_4$ is left behind. On the contrary, when pure $N_2$ is injected the front is much smoother, with the $N_2$ moving faster than $CH_4$ and overtaking it. As expected, injection of a mixture of $CO_2$ and $N_2$ results in the appearance of both the above mentioned effects. In particular, at the $CO_2$/$CH_4$ front the $N_2$ is enriched in the fluid phase, being the least adsorbing component.

Figure 6.5 shows the flow rates of $CO_2$, $CH_4$ and $N_2$ at the production well corresponding to the three different scenarios just described. It can be seen that when pure $CO_2$ is injected, the $CH_4$ recovery is completed as $CO_2$ breakthrough takes place, because of the characteristic displacement behavior described above. On the contrary, gas mixture containing $N_2$ shows an early breakthrough of the latter resulting in a produced stream of $CH_4$ polluted with $N_2$, until $CO_2$ breakthrough occurs. Moreover, a characteristic peak in the $CH_4$ production rate can be observed, which corresponds to the $N_2$ breakthrough (Bustin et al., 2008; Durucan and Shi, 2009).

Interestingly, the same trends presented above have been obtained by applying a completely different modeling approach, where the so-called local equilibrium assumption is made, i.e. adsorption and desorption occur quickly enough that the fluid phase and the coal matrix are always in equilibrium (Orr, 2007; Jessen et al., 2008). By neglecting dispersion phenomena, mass transfer resistance and swelling effects, the equations

# 6. Coal bed dynamics

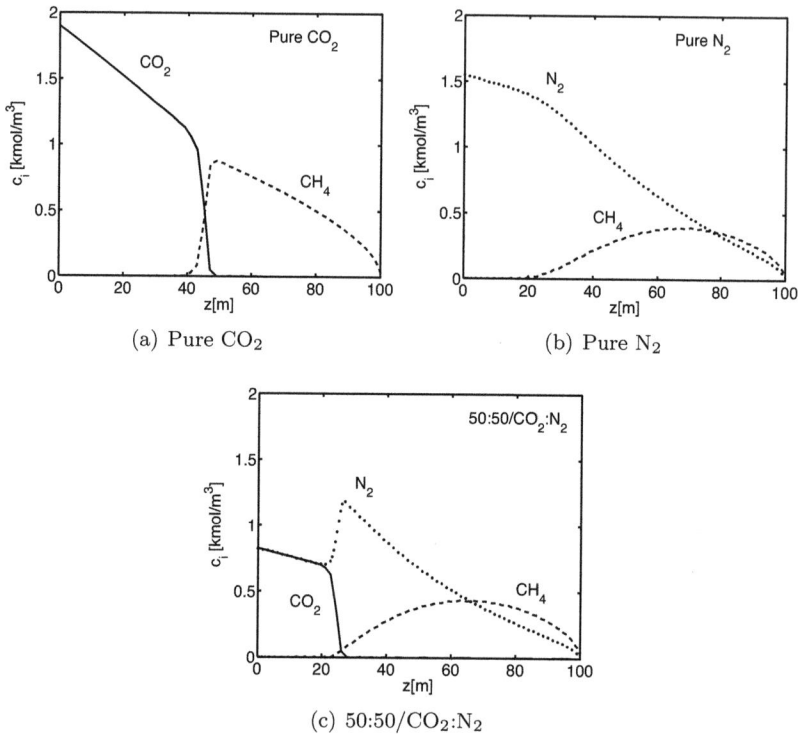

Figure 6.4: Concentration profiles of $CO_2$, $CH_4$ and $N_2$ along the coal bed axis for three different injection scenarios: pure $CO_2$ (a) pure $N_2$ (b) and 50:50/$CO_2$:$N_2$ (c).

## 6.3 Results

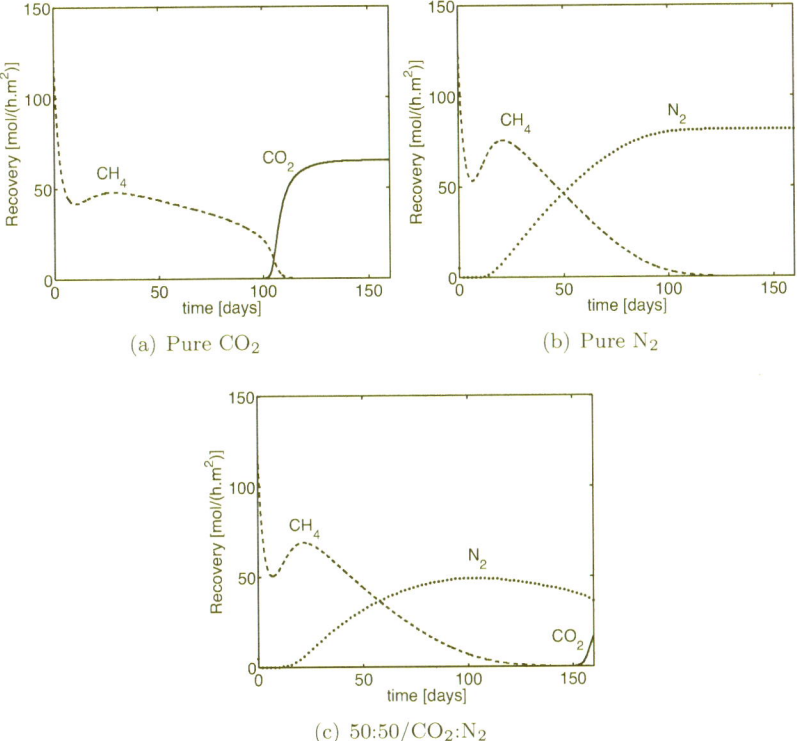

Figure 6.5: Flow rates of $CO_2$, $CH_4$ and $N_2$ at the production well as a function of time for three different injection scenarios: pure $CO_2$ (a) pure $N_2$ (b) and 50:50/$CO_2$:$N_2$ (c).

presented in Section 6.2 can be further simplified and a powerful mathematical technique, i.e. the method of characteristics, can be used to calculate the multicomponent flow in a coal bed. Even though it represents a strong simplification of the real coal seam, this model was able to describe the EBCM process in a way that sheds light on the complex injection/displacement dynamics. In particular, the $CO_2/CH_4$ displacement mentioned above has been described by the so-called "shock front", whereas the $N_2/CH_4$ through a much broader front, i.e. the so-called "simple wave" (Seto et al., 2006; Seto, 2007).

**Case A - weak swelling**

Figure 6.6 shows the produced gas quality (in terms of $CH_4$ purity) and the amount of $CH_4$ recovered as a function of time for different ECBM injection scenarios. It can be clearly seen that addition of $N_2$ into the injected gas results in increased pollution of the methane produced, due to overlap of the $N_2$ injection front and the $CH_4$ desorption fronts described previously. In the case of pure $CO_2$ in fact, the produced gas is almost pure $CH_4$ until completion of the recovery process. Moreover, in terms of amount of $CH_4$ recovered, injection of $N_2$-rich gas mixtures allows for a faster $CH_4$ recovery compared to the case where pure $CO_2$ is injected. However, in both figures it can be seen that the curve corresponding to the pure $CO_2$ injection crosses all the other curves and in particular the one corresponding to the mixture with composition 80:20/$CO_2$:$N_2$. This behavior can be explained by the more effective displacement of $CH_4$ by the $CO_2$ (due to its larger adsorptivity compared to both $CH_4$ and $N_2$), whose benefits are particularly evident when the swelling term, which affects the permeability, is relatively weak (Case A).
If one looks at the ECBM process first as a way to store $CO_2$, the $CH_4$ purity and the amount of $CH_4$ recovered should be plotted as a function

## 6.3 Results

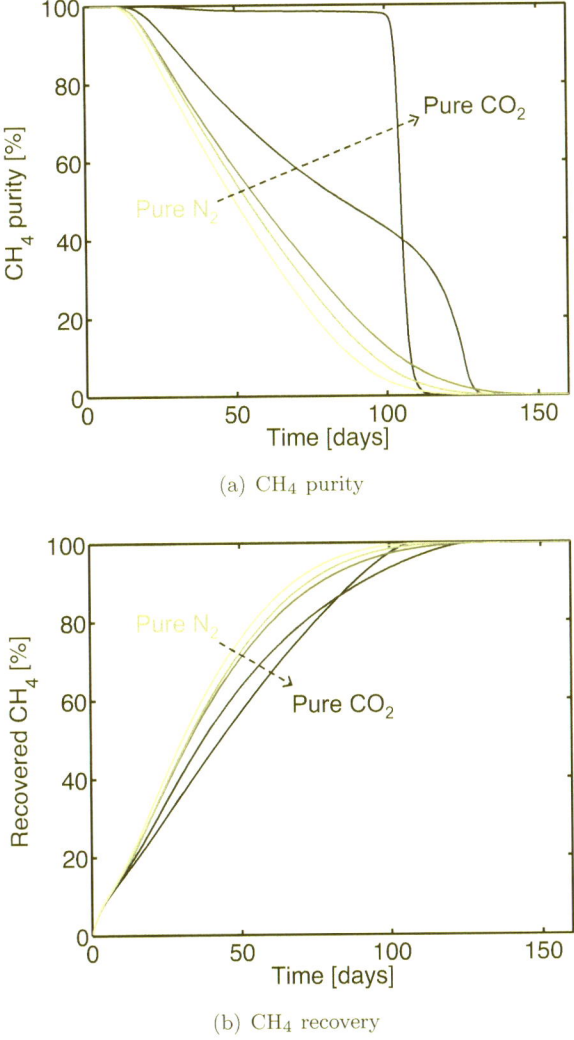

(a) CH$_4$ purity

(b) CH$_4$ recovery

Figure 6.6: CH$_4$ purity (a) and CH$_4$ recovery (b) as a function of time for different ECBM schemes with different injection compositions (Pure CO$_2$, 80:20/CO$_2$:N$_2$, 50:50/CO$_2$:N$_2$, 20:80/CO$_2$:N$_2$, pure N$_2$).

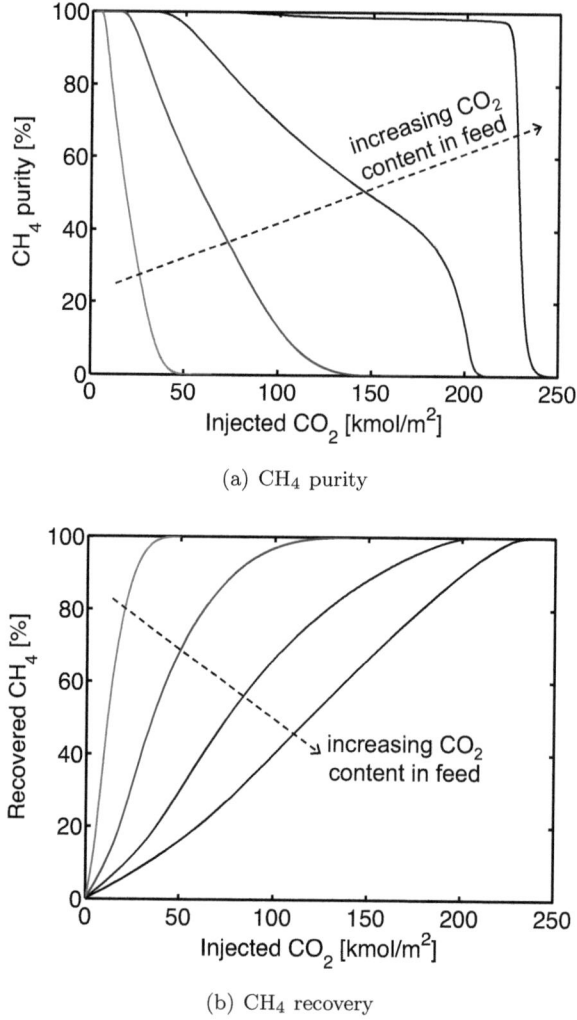

(a) CH$_4$ purity

(b) CH$_4$ recovery

Figure 6.7: CH$_4$ purity (a) and CH$_4$ recovery (b) as a function of the amount of CO$_2$ injected for different ECBM schemes with different injection compositions (Pure CO$_2$, 80:20/CO$_2$:N$_2$, 50:50/CO$_2$:N$_2$, 20:80/CO$_2$:N$_2$).

## 6.3 Results

of the amount of $CO_2$ injected instead of time, as shown in Figure 6.7. As expected, in this case the intersection disappears. Two comments are worth making with respect to these outcomes. On the one hand, if one is interested in the recovered methane as a fuel or a technical gas, there is a clear trade-off between the incremental methane recovery that can be achieved and the produced gas quality. On the other hand, if the goal is that of storing $CO_2$ that has been captured, then the amount of $CO_2$ that can be injected and stored in the reservoir is of primary importance. Figure 6.8 shows the amount of $CO_2$ injected and stored in the coal bed as a function of time for the different ECBM schemes. It can be seen that in both figures the obtained curves are clearly fanning out, with the consequence that $CO_2$ rich mixtures should be preferred if the goal is to maximize $CO_2$ storage. Again, this is particularly true for the situation investigated here (Case A) where swelling effects which may reduce the permeability and therefore the injectivity are relatively small.

### Case B - strong swelling

The same situation as the previous section is investigated, with the difference that the swelling effect in the permeability relationship is now greater. Figure 6.9 shows the produced gas quality (in terms of $CH_4$ purity) and the amount of $CH_4$ recovered as a function of time for different ECBM injection scenarios. Qualitatively the same conclusion as for Figure 6.6 can be drawn, with an improved methane recovery coupled to a deterioration of the produced $CH_4$ purity, resulting from a $N_2$-enrichment of the injected gas mixture. However, compared to the situation of Case A, the curves are much more spread and the difference between pure $CO_2$ and pure $N_2$ injection more evident. Particularly

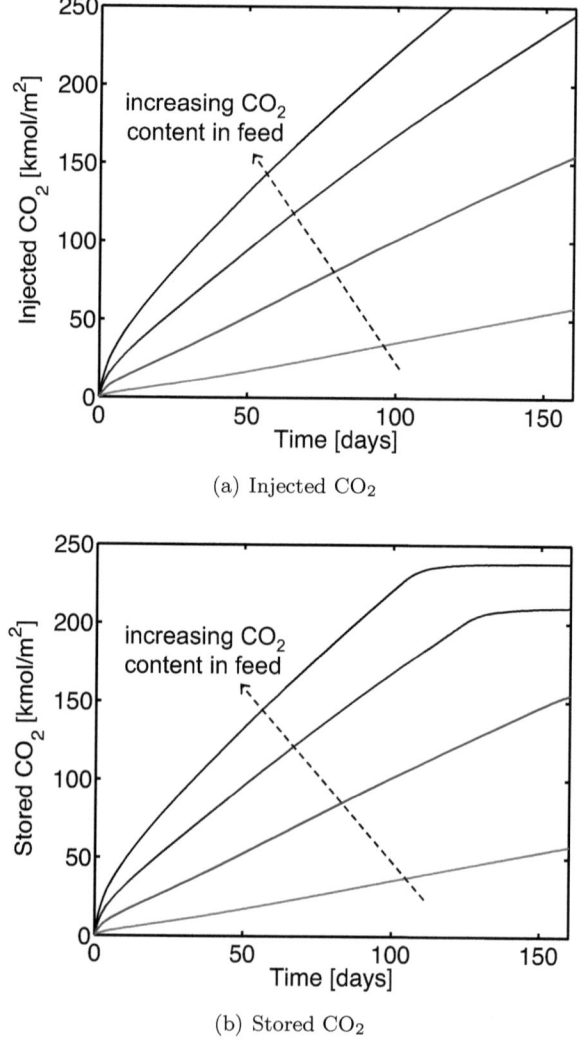

(a) Injected $CO_2$

(b) Stored $CO_2$

Figure 6.8: Amount of $CO_2$ injected (a) and stored (b) as a function of time for different ECBM schemes with different injection compositions (Pure $CO_2$, 80:20/$CO_2$:$N_2$, 50:50/$CO_2$:$N_2$, 20:80/$CO_2$:$N_2$).

## 6.3 Results

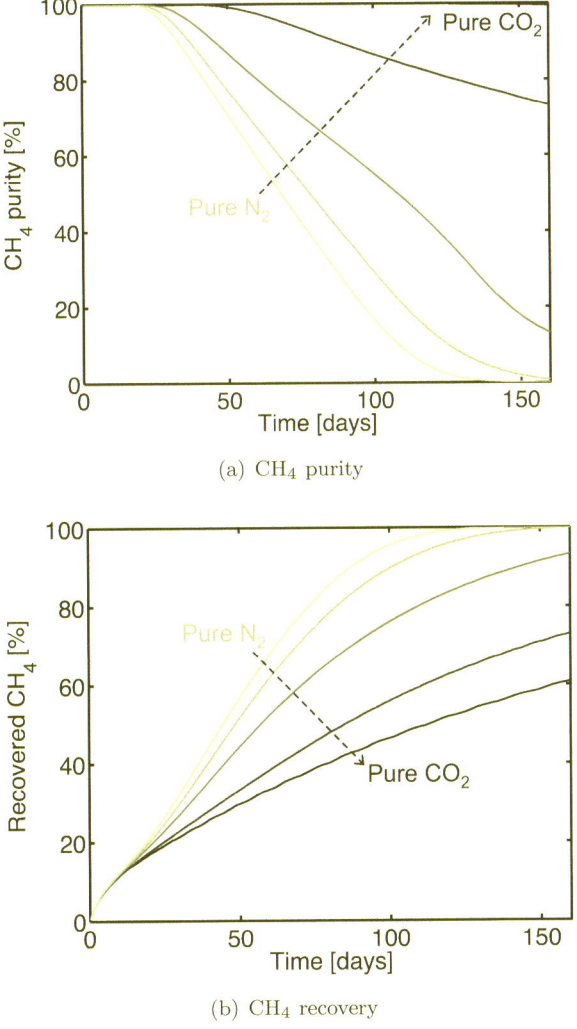

(a) CH$_4$ purity

(b) CH$_4$ recovery

Figure 6.9: CH$_4$ purity (a) and CH$_4$ recovery (b) as a function of time for different ECBM schemes with different injection compositions (Pure CO$_2$, 80:20/CO$_2$:N$_2$, 50:50/CO$_2$:N$_2$, 20:80/CO$_2$:N$_2$, pure N$_2$).

interesting is the outcome of Figure 6.10, where the amount of $CO_2$ injected and stored in the coal bed is plotted as a function of time for the different ECBM schemes. It can be seen that, due to the loss in injectivity caused by the strong reduction in permeability (see Figure 6.3), the difference between the different injection scenarios is much smaller compared to Case A. In particular, the amount of $CO_2$ injected in the pure $CO_2$ case is only slightly higher than the 80:20/$CO_2$:$N_2$ case. It is easy to imagine, that the two curves may eventually cross, with the surprising result that a gas mixture with a lower $CO_2$ content would allow injecting (and therefore storing) more $CO_2$ compared to the case where pure $CO_2$ is injected. An extreme scenario would correspond to the complete blocking of coal fractures, impeding therefore to exploit the whole coal bed volume available. Differently from Case A, in this situation it would be more efficient to inject $CO_2$/$N_2$ mixtures instead of pure $CO_2$ even if the objective of the ECBM project is to maximize $CO_2$ storage. It is worth noting that, only an injection scheme has been considered here, i.e. 1 injection and 1 production well, but that similar results have been obtained for others multi-well patterns (Bustin et al., 2008; Durucan and Shi, 2009)

## 6.4 Discussion and concluding remarks

In this study, the gas flow dynamics during an ECBM operation have been studied with the help of a one-dimensional mathematical model, consisting of mass balances accounting for gas flow and sorption, and a constitutive stress-strain relationship for the description of porosity and permeability changes during injection, which has been validated in a previous work (Pini et al., 2009). The model allowed us to highlight some important aspects playing an important role in controlling the gas flow

## 6.4 Discussion and concluding remarks

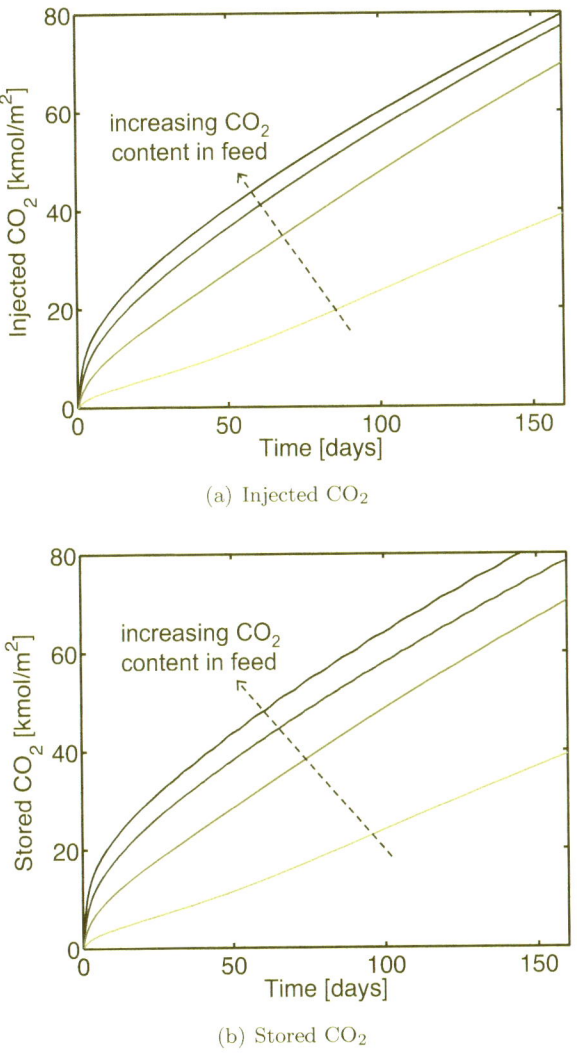

(a) Injected $CO_2$

(b) Stored $CO_2$

Figure 6.10: Amount of $CO_2$ injected (a) and stored (b) as a function of time for different ECBM schemes with different injection compositions (Pure $CO_2$, 80:20/$CO_2$:$N_2$, 50:50/$CO_2$:$N_2$, 20:80/$CO_2$:$N_2$, pure $N_2$).

dynamics during the injection process. In particular, it was shown that gas injection can indeed enhanced methane recovery. Moreover, $CO_2$ injection yielded a reduction in permeability and, in the cases where $N_2$ was injected, a much more rapid response coupled with an earlier breakthrough of $N_2$ was observed. The former was attributed to the closing of the fractures associated with coal swelling, particularly evident near the well, where the $CO_2$ concentration is at the highest. In the case of $N_2$ coal undergoes a net shrinking, but the $N_2$ injection front and the $CH_4$ desorption front overlap so as the injected gas pollutes methane much more than in the case of $CO_2$. These considerations are in agreement with the observations from ECBM field studies.

The 1-D description presented above provide a very useful understanding of the key mechanisms that affect the storage and recovery process. However, a description which is much closer to reality is obtained with reservoir modeling, where 1-D model presented in this study can be incorporated and extended to a 3-D multi-phase (coal, gas and water) model. Such models have to be solved in a 3-D domain that comprises the coal bed and account for its geological structure and possibly heterogeneous physical features as well as for the configuration of the injection and production wells. The data obtained from the field tests are the information needed to be used for the calibration of these reservoir simulators (history-matching). Once validated, these models represent important tools to design ECBM processes. In fact, only a throughout understanding of the different mechanisms acting during ECBM allows to critically assess the success or the failure of the performed field tests and to plan future demonstration projects. This very important approach has been applied to the data of the South Qinshui Basin single well micro-pilot test, where after successful history matching a larger scale multi-well field test has been designed and planned using the simu-

## 6.4 Discussion and concluding remarks

lator (Shi et al., 2008). Similarly, a reservoir simulator has been used in support of the operations of the Coal-Seq project (Reeves, 2004). Durucan and Shi have consistently and successfully used the Imperial College in-house ECBM simulator METSIM2 to history match field data from the Coal-Seq project (Allison unit) (Shi and Durucan, 2004b), from the Alberta project (flue gas injection) (Shi and Durucan, 2005b), and from the JCOP project for both single and multi-well tests (Shi et al., 2008). The PSU-COALCOMP reservoir simulator developed at The Pennsylvania State University has been used to study the effect of the well configuration and design on the stored amount of $CO_2$ as compared to its theoretical amount, that is given in principle by the sorption isotherm (Bromhal et al., 2005; Sams et al., 2005). It was shown that depending on the sorption time constant used, a the end of the project life time a significant portion of the swept region can be still far from equilibrium, resulting in a reduced amount of $CO_2$ stored, i.e. down to 50% as compared to the thermodynamic limit predicted by the adsorption isotherm. By investigating different well configurations this situation can be improved and useful design criteria can be derived.

In conclusion, the design of gas injection and coal bed methane displacement is an important area of research aiming at optimizing the economics and the effectiveness of the ECBM process as a whole, in terms of injectivity, amount of $CO_2$ stored and amount and purity of $CH_4$ produced. The results of the present study suggest that there is space for optimizing the ECBM process depending on whether the objective of the project is to maximize the $CO_2$ storage or the methane recovery. In addition to that, operational constraints, such as the composition of the injected gas stream or the required purity of the produced gas depending on its future use, would also play an important role in the design process. Based on the observation from this and other studies, it would be extremely attractive to inject flue gas directly without the expensive capture step and as a way to keep permeability sufficiently high. Answering whether this is indeed possible, in spite of the increased gas compression costs due to the additional nitrogen, requires more research work and field test under different conditions and geological settings. Efforts in this direction are justified by the ultimate goal of increasing the confidence in ECBM process as a way for reducing greenhouse gases emissions.

# Chapter 7

# Containment in the reservoir

## 7.1 Introduction

When $CO_2$ is injected in a coal seam it is trapped by the mechanism of adsorption on the coal surface. As explained in Chapter 2, this process is controlled by a thermodynamic equilibrium between the amount of $CO_2$ which is adsorbed on the coal surface and its corresponding density (or pressure) in the fluid phase. This relation is described by the adsorption isotherm and implies that $CO_2$ will stay in the adsorbed state as long as the pressure (or density) of the fluid phase in the seam is maintained. On the one hand, at the conditions where $CO_2$ storage is supposed to be feasible, the injected $CO_2$ is lighter than water and therefore would tend to migrate upward through the porous overlying strata. On the other hand, this buoyancy-driven migration is prevented and controlled by several mechanisms, which are described in the following.

First, there are naturally occurring geometrical configurations of permeable and impermeable layers, i.e. the so-called traps, that allows for an effective accumulation of fluids. The presence of a low permeability rock, i.e. the so-called cap rock, which surrounds the coal seam and therefore impedes the upward and lateral movement of the injected fluid is part of this concept. With reference to Figure 7.1, one can distinguish between two kind of traps: structural traps, where the sealing results from a tectonic movement (folds, A and faults, B), and stratigraphic traps (C), which are the result of local variation in the mechanism of sediments deposition, thus forming a porous area included in impermeable strata (Bachu, 2008). As an example, in Northern Switzerland, a combination of compressional and slip movements formed the so-called Permo-Carboniferous trough, a structural depression bounded by faults (Diebold, 1988; Nagra, 2002). Thanks to this existing structural trap, the numerous coal seams present in this trough could have retained methane and/or be suitable for $CO_2$ storage purposes. These containment mechanisms are often referred to as primary mechanisms, because their acting is essential since the beginning of the injection process.

Secondly, since the caprock is naturally filled with water, capillary phenomena are present, which can inhibit volume flow (Li et al., 2006). The capillary pressure $P_c$ is defined as the pressure difference between the non-wetting phase ($CO_2$, at a pressure $P_g$) and the wetting phase (water or brine, at a pressure $P_w$). The gas will enter a pore filled with water only if their pressure difference exceeds the capillary pressure, as shown in Figure 7.2a. Therefore, as long as this condition is not met, flow will not be possible, even if a pathway were available (Berg, 1975). The value of the capillary pressure depends on the physical properties of both rock and fluid. In particular, its value is inversely proportional to

## 7.1 Introduction

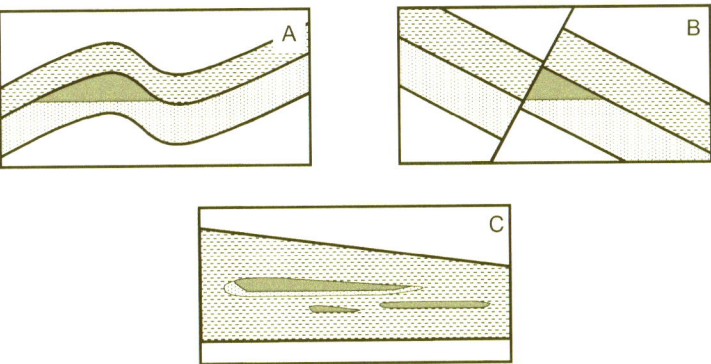

Figure 7.1: Schematic of the three main trapping configurations: structural traps, i.e. folds (A) and faults (B) and stratigraphic traps (C).

the pore radius (Berg, 1975), and for rocks with very small pores, it can reach extremely high values. In Figure 7.2b, the capillary pressure $P_c$ calculated from the well known Young-Laplace equation is shown as a function of the pore radius. It can be seen that it is negligible for larger radii, i.e. $r > 0.5$ $\mu$m, but it increases as the rock radius decreases. As an example, for the Midale Evaporite, the seal rock of the Weyburn Field in Canada, where $CO_2$ is being injected into an oil reservoir, a value of about 210 bar has been measured (Li et al., 2005), corresponding to a radius $r < 0.005$ $\mu$m. This value is so high, that capillary failure is almost inconceivable and leakage could only occur through fracturing of the caprock (Watts, 1987).

Under certain circumstances, seal failure may happen and the injected $CO_2$ would therefore start displacing the water that occupies the pore space of overlying rock strata in a drainage-like mechanism. However, as the $CO_2$ plume moves upwards, other trapping mechanisms may come into play. In fact, the water cannot be completely displaced by the $CO_2$,

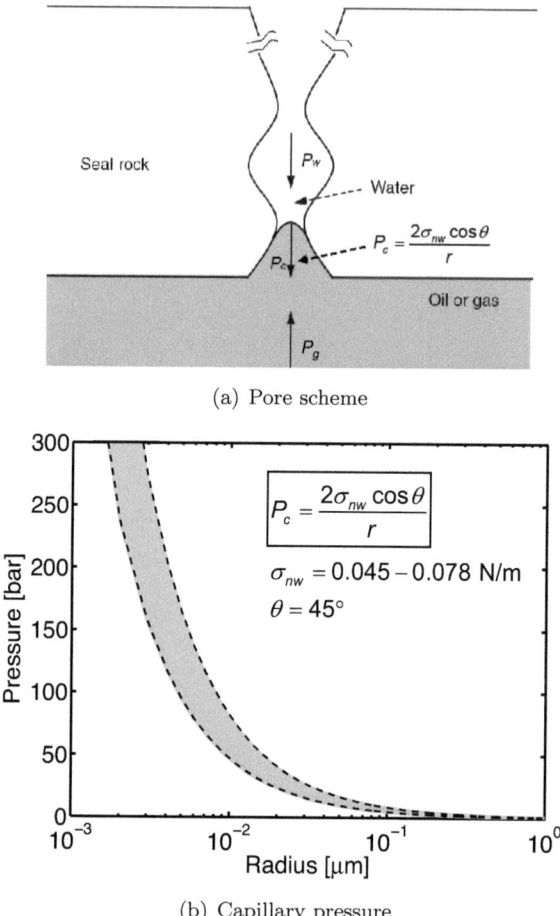

(a) Pore scheme

(b) Capillary pressure

Figure 7.2: Trapping by capillary phenomena. (a) Schematic of capillary sealing mechanism in a pore throat of a seal rock. Modified from Li et al. (2005). (b) Capillary pressure curves calculated with the Young-Laplace equations for a $CO_2$-water system, with wetting angle $\theta = 45°$ and interfacial tension $\sigma_{nw}$ ranging from 0.045 and 0.078 N/m, as reported by Hildenbrand et al. (2004).

## 7.1 Introduction

but remains present in small pores that have not been reached by the $CO_2$ as well as along the edges of the occupied pores (Lenormand et al., 1983). This phenomenon makes it possible that at the trailing edge of the $CO_2$ plume, water re-invades the pore space (imbibition). The mechanism dominating this process is the the so-called snap-off, where the increase of the water saturation leads to the by-passing and, consequently, to the disconnection of $CO_2$ bubbles (Lenormand et al., 1983; Juanes et al., 2006). This phenomenon is referred to as residual trapping, since these $CO_2$ bubbles left behind the moving plume are effectively immobile and therefore permanently trapped (Bachu, 2008). For instance, it has been shown that this mechanism is an essential contribution to $CO_2$ storage in saline aquifers (Juanes et al., 2006; Hesse et al., 2008).
Finally, over the time of the storage process $CO_2$ may dissolve into the formation water. First, the dissolved $CO_2$ is trapped since water or brine saturated with $CO_2$ are heavier than water and will therefore migrate downward. Secondly, the dissolved $CO_2$ may be involved in chemical reactions with the rock matrix (mineral trapping). As a result, the $CO_2$ would be permanently fixed as a carbonate (Gunter et al., 1993; Bachu et al., 1994).
All the different mechanisms mentioned above (structural, capillary, residual and mineral trapping) act from the beginning of the storage operation, but their contribution to the storage of $CO_2$ changes over time. Structural trapping is essential during the operational phase, whereas the contribution of the much slower capillary, residual and mineral trapping (the so-called secondary trapping mechanisms) increases in the post-injection phase. With the grow up of the secondary trapping mechanisms, the storage security increases as well (Bachu, 2008). In fact, the amount of $CO_2$ that may migrate out of the reservoir diminish over time since, at the stage where the contribution of the secondary mechanism

become significant, a corresponding significant amount of $CO_2$ is immobile in the pore space or fixed in carbonates.

The situation of a leakage from a coal seam, but also from any other reservoir, has to be taken into consideration in the assessment of the storage potential of the given reservoir, in particular for safety issues. In such a scenario, the storage of $CO_2$ in the underground reservoir can be viewed as an injection process in the overlying rock strata. The mechanisms controlling the $CO_2$ flow can be divided into two main categories, namely diffusive and convection-driven flow (Schlomer and Krooss, 1997). A comprehensive description of the leaking process involves these two flow and transport processes, and has to be solved by 3D multiphase, multi-component models accounting for the geological structure of the reservoir and its heterogeneous physical features. As a first step towards a better understanding of this flow behavior, in the present work the diffusive and convective flow are studied independently as they can be seen as the two limiting cases of the real flow pattern. In particular, one-dimensional models are presented for both flow regimes, in order to identify and quantify their time-scale (Section 7.2). It should be pointed out, that even if these models assume a strong simplification of the real geological situation, they will allow us to bring important conceptual insights concerning the leakage process from the reservoir. The second part of this chapter focuses on coal seams, and in particular on the interpretation of the gas-in-place (GIP), a quantity that allows to obtain information on the sealing efficiency of the caprock overlying the coal seam (Section 7.3).

## 7.2 Modeling the flow through the caprock

### 7.2.1 Diffusion

As mentioned above, diffusion is one of the two mechanisms, which control the flow of $CO_2$ out of the coal seam and through the cap rock. The strata surrounding the coal seam, including the cap rock, are naturally saturated with water, where $CO_2$ can be dissolved. This gives rise to the diffusive flow, a perpetual and ubiquitous migration process through the water occupying the pore space and driven by the concentration gradient (Schlomer and Krooss, 1997). Due to its nature, it has been shown that diffusion may be a quantitatively significant process over a geological time scale (Montel et al., 1993).

A one dimensional model has been developed to describe the $CO_2$ losses due to diffusion from a coal seam. When describing diffusion in liquid filled pores, it is common to define a pore diffusivity $D_p$, under the assumption that flux occurs only in these pores and the one through the solid can be neglected (Ruthven, 1984). This new defined pore diffusivity differs from the conventional one, in the fact that it accounts for the random orientation of the pores and the variation in their diameter. Both effects give rise to a longer diffusion path and are described by a tortuosity factor $\tau$, i.e. $D_p = D/\tau$. The flux across a porous surface filled with water can be therefore written as

$$J = -\varepsilon D_p \frac{\partial c}{\partial z} \quad (7.1)$$

where $c$ is the solute concentration and $\varepsilon$ the porosity. Based on Fick's second law, the governing transport equation is thus given by

$$\varepsilon \frac{\partial c}{\partial t} = \frac{\partial}{\partial z}\left(\varepsilon D_p \frac{\partial c}{\partial z}\right) \quad (7.2)$$

This equation can be solved by giving an initial condition and two boundary conditions, corresponding to the contact interface between the coal seam and the cap rock and the upper boundary of the cap rock, respectively.

## 7.2.2 Convection

During the ECBM operation, the $CO_2$ pressures in the coal seam can be higher than the natural (hydrostatic) one, especially close to the injection well. Under these conditions it can happen that the capillary pressure is exceeded and $CO_2$ starts moving in gaseous form. In addition, during the diffusion process described above, gas concentration may reach the local saturation point, causing free gas to appear. Both phenomena give rise to a multi-phase flow (gas/water), where the $CO_2$ migration is controlled by the relative permeability of the gas phase and of the water phase. Similarly to the case where $CO_2$ is injected into saline aquifers, the non-wetting $CO_2$ gas phase invades the pore space in a drainage-like process, thus displacing the water.

In order to describe the water displacement by the injected $CO_2$, a one dimensional model has been derived. Under the assumption that the two fluids ($CO_2$ and $H_2O$) are immiscible, a material balance can be written for each phase:

$$\varepsilon \frac{\partial}{\partial t}(\rho_g S) + \frac{\partial}{\partial z}(\rho_g u_g) = 0 \qquad (7.3a)$$

$$\varepsilon \frac{\partial}{\partial t}[\rho_w (1-S)] + \frac{\partial}{\partial z}(\rho_w u_w) = 0 \qquad (7.3b)$$

where $S$ is the gas saturation, defined as the ratio between the volume occupied by the gas phase and the total pore volume, i.e. $V_g/V_p$; $\rho_g$

## 7.2 Modeling the flow through the caprock

and $\rho_w$ are the gas and water phase molar densities, $u_g$ and $u_w$ are the superficial velocity in the gas and water phase, $t$ is the time and $z$ the axial coordinate. By further assuming that the water density $\rho_w$ remains constant and rearranging, the following system of partial differential equations in the two unknowns $S$ and $\rho_g$ is obtained:

$$\frac{\partial S}{\partial t} = \frac{1}{\varepsilon}\frac{\partial u_w}{\partial z} \tag{7.4a}$$

$$\frac{\partial \rho_g}{\partial t} = -\frac{\rho_g}{\varepsilon S}\left(\frac{\partial u_w}{\partial z} + \frac{\partial u_g}{\partial z}\right) - \frac{u_g}{\varepsilon S}\frac{\partial \rho_g}{\partial z} \tag{7.4b}$$

Darcy's law is used to describe the convective flow, and in each phase the flow velocities are given by

$$u_g = -\frac{k_g}{\mu_g}\left(\frac{\partial P_g}{\partial z} + \rho_{m,g}g\right) \qquad u_w = -\frac{k_w}{\mu_w}\left(\frac{\partial P_w}{\partial z} + \rho_{m,w}g\right) \tag{7.5}$$

where $k_j$, $\mu_j$, $P_j$ and $\rho_{m,j}$ are respectively the effective permeability, the viscosity, the pressure and the mass density in phase $j$ ($=$ g or w). The second term in the brackets accounts for the gravitational effects, which, being the flow vertical, counteracts the upward migration. Note that the (absolute) permeability $k$ is a characteristic parameter of the porous medium, which is independent of the flowing fluid. In the case of multi-phase flow, relative permeabilities are commonly used, which are defined as:

$$k_{rg} = \frac{k_g}{k} \qquad k_{rw} = \frac{k_w}{k} \tag{7.6}$$

Simple relationships are used to relate the relative permeability to the gas saturation $S$ for both the gas and water phase (Orr, 2007):

$$k_w = k(1-S)^2 \qquad k_g = S^2 \qquad (7.7)$$

As explained above, the pressures are different in the two phases, as they must be due to capillary effects, and they are related by the capillary pressure as:

$$P_c = P_g - P_w \qquad (7.8)$$

where $P_c$ is the capillary pressure, which is again a function of the porous medium and the fluids involved. In fact, looking at a single pore, the capillary pressure can be expressed with the well-known Young-Laplace equation:

$$P_c = \frac{2\sigma_{gw}\cos\theta}{r} \qquad (7.9)$$

where $\sigma_{gw}$ is the $CO_2$/water interfacial tension, $\theta$ is the contact angle between the $CO_2$/water separation surface and the solid and $r$ the radius of the pore.

Finally, the Span and Wagner equation of state is used to relate pressure and density of the $CO_2$ phase (Span and Wagner, 1996). The model is completed by giving an initial conditions for the two unknowns $\rho_g$ and $S$ and boundary conditions at the lower and upper boundary of the cap rock.

### 7.2.3 Solution Procedure

In order to compare the two mechanisms of flow, the two corresponding models are solved for a similar geological situation. In particular, a coal

## 7.2 Modeling the flow through the caprock

seam is considered which lies at 1500 m below a cap rock layer with 3% porosity and an arbitrary thickness of 1 m. At this depth, typical conditions encountered in terms of pressure and temperature are 150 bar and 75°C (see Chapter 1). It is further assumed that $CO_2$ has been injected into the coal seam, and that at the time when the simulation starts the $CO_2$ pressure in the coal seam is equal to the hydrostatic pressure (diffusion scenario) or slightly above (convection scenario). Moreover, the cap rock is initially completely filled with water and almost free of $CO_2$. In the case of diffusion, the initial $CO_2$ concentration in the cap rock is $c_0$=0.03 g/L. The concentration of $CO_2$ at the coal seam/caprock boundary is assumed to be constant over the time of the simulation and takes the value of about 46 g$CO_2$/L. This value corresponds to the equilibrium concentration of a water solution in contact with a $CO_2$ phase at 150 bar and 75°C (Spycher et al., 2003). At the upper boundary of the cap rock, $CO_2$ concentration is kept constant at a value of 0.03 g/L, i.e. equal to the initial concentration $c_0$.

In the case of convective flow, the cap rock is initially saturated with water, i.e $S \approx 0$, with a density of 986 g/L and a pressure of 150 bar. At the coal seam/caprock boundary only $CO_2$ is present ($S = 1$) and an overpressure of 20 bar is imposed. Since the objective of the simulations is to obtain an estimate of the breakthrough time of the $CO_2$ at the cap rock upper boundary once gas flow has started, a relatively large value of the pores size in the cap rock has been chosen, i.e. $r = 1$ μm. This allows keeping the capillary pressure, and therefore the resistance to gas flow, low. Moreover, only this type of pores is present in the cap rock and as a consequence one value only for the capillary pressure is obtained; according to Eq.(7.9) and by taking an interfacial tension of 0.062 N/m and a wetting angle of 45°, it takes the value of about 0.9 bar. Finally, at the upper boundary of the cap rock, the capillary pressure disappears, and the gas pressure is kept constant at the value of the hydrostatic water pressure. Finally, values for the viscosity have been obtained from the National Institute of Standards and Technology (NIST) and for the $CO_2$ and $H_2O$ phase at the temperature and pressure conditions of the simulations take the value of $3.9 \cdot 10^{-5}$ Pa.s and $4.0 \cdot 10^{-4}$ Pa.s, respectively.

All the parameters used for the model evaluation are listed in Table 7.1.

Table 7.1: Parameter used for the model evaluation.

| Property | Value |
|---|---|
| $\varepsilon$ | 0.03 |
| $L$ [m] | 1 |
| $T$ [°C] | 75 |
| **Diffusion model** | |
| $D$ [m²/s] | $1 \times 10^{-9}$ |
| $\tau$ | 3 |
| $c_0$ [g/L] | 0.03 |
| $c(z=0\text{ m})$ [g/L] | 46 |
| $c(z=1\text{ m})$ [g/L] | 0.03 |
| **Convective model** | |
| $\theta$ [°] | 45 |
| $k$ [mD] | $1 \times 10^{-6}$ |
| $\mu_g$ [Pa·s] | $3.9 \times 10^{-5}$ |
| $\mu_w$ [Pa·s] | $4.0 \times 10^{-4}$ |
| $P_{w0}$ [bar] | 150 |
| $P(z=0\text{ m})$ [bar] | 170 |
| $r$ [m] | $10 \times 10^{-6}$ |
| $\sigma_{gw}$ [N/m] | 0.062 |
| $S_0$ | 0.01 |
| $S(z=0\text{ m})$ | 1 |

## 7.2 Modeling the flow through the caprock

### 7.2.4 Results

Figure 7.3 shows the concentration profiles in the cap rock at different time intervals for the diffusion case. Note that the concentrations have been normalized by the constant value imposed at the coal seam/caprock boundary, i.e. at $z=0$ m. The profiles show the usual smooth decay of the diffusive flow and it can be seen that, for a 1 m thick cap rock, breakthrough occurs after about 5 years. Results are shown for the first 100 years only, since from that point on steady state has been reached and the profiles don't change anymore. The cumulative flux obtained at the upper cap rock boundary obtained for a longer time horizon is shown in Figure 7.4 for three different diffusion coefficients. For all the plotted curves the observed behavior is qualitatively the same and, as expected, breakthrough time increases with decreasing diffusion coefficient: from about 1 year up to 100 years.

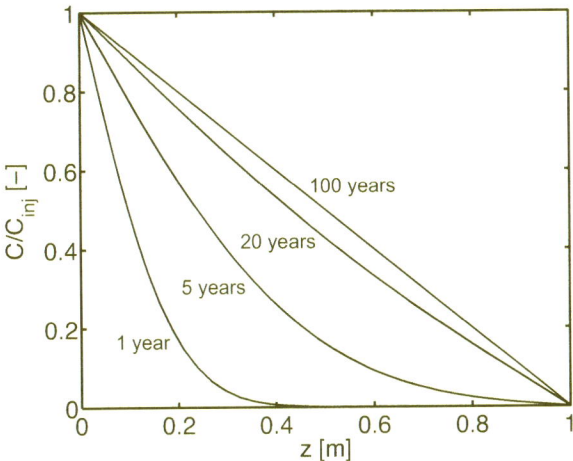

Figure 7.3: $CO_2$ concentration profiles in the 1 m thick caprock at different time intervals. Diffusion coefficient: $D=1\times10^{-9}$ m$^2$/s.

# 7. Containment in the reservoir

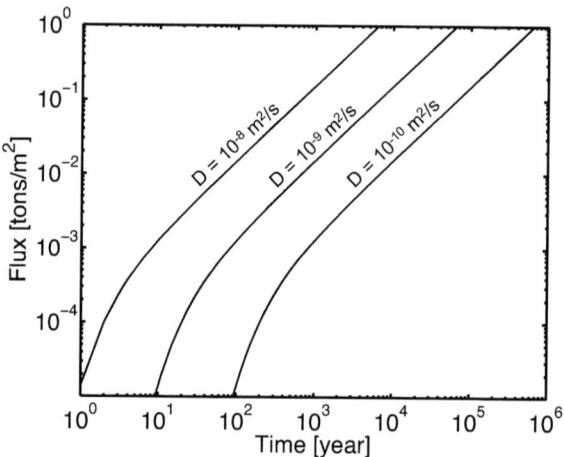

Figure 7.4: $CO_2$ cumulative flux obtained at the upper caprock boundary, i.e. at $z=1$ m, for three different diffusion coefficients.

Similar plots have been prepared for the convection case, i.e. when $CO_2$ migrates as a separate phase through the caprock. Figure 7.5 shows the gas saturation $S$ along the cap rock at different time intervals. It can be seen that gas invades the pore space as time advances with a drainage-like mechanism. As long as gas breakthrough does not occur, the saturation profiles are characterized by a smooth decay, which flattens out at a $S$ value of about 0.45 and followed by a sharp front. A similar behavior has been reported for water flooding of oil reservoirs, where the water/oil displacement can be described by the sequence of two characteristic propagation fronts, the so-called simple wave followed by a semi-shock (Rhee et al., 2001). Once that the gas phase has reached the upper boundary of the cap rock, it starts filling more and more the pore space, reaching an average saturation of about 90% after 460 years. The cumulative flux of the $CO_2$ leaving the cap rock is shown as a function of time in Figure 7.6 for four different permeabilities, i.e. from $10^{-3}$

## 7.2 Modeling the flow through the caprock

down to $10^{-6}$ mD. It is worth pointing out that the latter corresponds to the permeability of the Opalinus Clay, a low-permeability sedimentary rock in Northern Switzerland identified as a potential host formation for radioactive waste disposal (Marschall et al., 2005). All the plotted curves are qualitatively identical, characterized by an initial sharp rise corresponding to the characteristic gas saturation profiles described above. As expected, with decreasing permeability, breakthrough time increases. For the sake of comparison, the flux obtained for a diffusive flow scenario (dashed line) is shown in Figure 7.6 together with the outcomes obtained for the convective flow. It can be seen that, even if in one case, i.e. $k=10^{-6}$ mD, breakthrough for the convective flow occurs later than for the diffusive flow, the capacity of the former exceeds by several orders of magnitude that of the latter. In this context therefore, the convective flow has to be seen as the most dangerous mechanism for gas leakage from a storage reservoir.

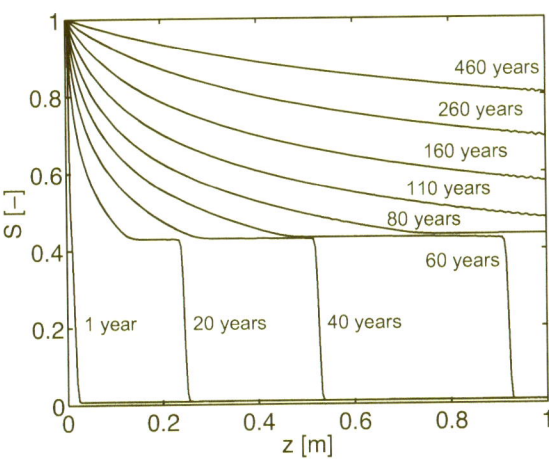

Figure 7.5: Gas saturation profiles in the 1 m thick cap rock at different time intervals. Cap rock permeability $k=1\times 10^{-6}$ mD.

7. Containment in the reservoir

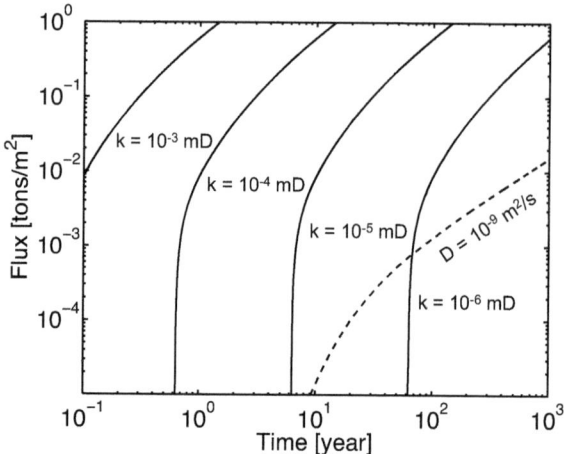

Figure 7.6: $CO_2$ cumulative flux obtained at the upper cap rock boundary, i.e. at $z=1$ m, for four different permeability. Diffusive flux is shown for the sake of comparison (dashed line).

### 7.2.5 Discussion

All the results described above have been performed by taking an arbitrary thickness of the cap rock of 1 m, since the objective was to show the main characteristics of the two flow mechanisms and to compare them. Cap rocks can in fact be thicker: for example, the cap rock succession overlying the Utsira reservoir at Sleipner is several hundred meters thick (Chadwick et al., 2004), whereas the Midale Evaporite, the seal rock of the Weyburn field in Canada (EOR), is 2-11 m thick (Li et al., 2005). On the one hand, we have seen that the most important parameter controlling the capillary sealing mechanism, i.e. the one impeding the dangerous leaking process driven by convective flow, is the pore radius of the cap rock. As long as the difference between the $CO_2$ and water pressure doesn't exceed the capillary pressure of the cap rock, its thickness is less relevant, in accordance with other studies (Watts, 1987). In this work, capillary pressure was kept small on purpose, allowing us to study the characteristic flow pattern once convective flow has started. On the other hand, a thicker cap rock would delay the breakthrough time in the case of seal failure; this is an important consideration as long as one considers gas to be stored, and therefore separated from the atmosphere, if it has left the reservoir, but it is still in the cap rock and on the way towards the surface. In this context, residual trapping, which was not considered in the calculation presented above, would play an essential role in the immobilization of $CO_2$ bubbles in the pore space of the rocks, as for the case of $CO_2$ storage in saline aquifers (Juanes et al., 2006; Hesse et al., 2008). Finally, it is worth mentioning again that a detailed quantitative analysis should take into account the 3D geological structure of the reservoir/seal pair and possibly the heterogenous physical features of the layers constituting it, such as porosity and permeability.

## 7.3 The gas-in-place in coal seams

A unique feature of coal seams is that they are both a source rock, where methane is generated, and a reservoir, where methane or other gases can be accumulated (Levine, 1993). Differently to conventional gas reservoirs, porosity of coal seams is very low, i.e. at maximum 5%, and storage is provided by the ability of coal to sorb gases (see Chapter 2). In order to better clarify this aspect, Figure 7.7 shows the amount of $CO_2$ stored as a function of the pressure, for three different reservoirs of same size (100 × 100 × 2 m), namely an aquifer with 25% porosity and two coal seams with the same porosity (5%), but different sorption capacities. The latter have been chosen to be representative for the range

# 7. Containment in the reservoir

of adsorption results reported in Chapter 2. It is worth pointing out that the reservoirs are assumed to be dry, i.e. the $CO_2$ exploits all the pore space available in the case of an aquifer and has the surface area for adsorption at its full disposal in the case coal seams. It can be seen that although coal seam porosity is very small compared to aquifers, the amount of $CO_2$ which can be stored in the reservoirs is similar. Moreover, it is evident from the figure that adsorption shows its benefits in terms of storage capacity in the low pressure range, whereas the bigger pore space available in the aquifer becomes more important as the pressure increases, i.e when $CO_2$ become denser.

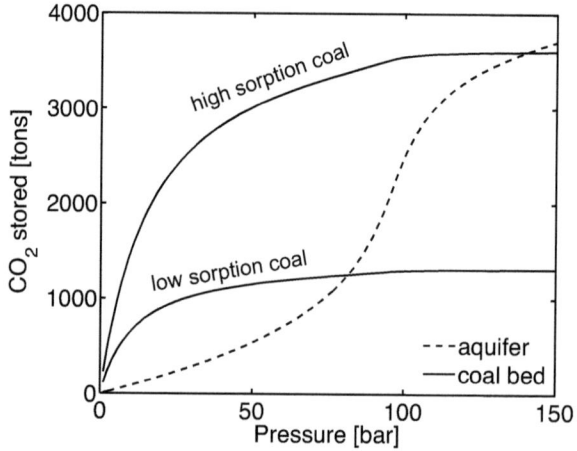

Figure 7.7: Amount of $CO_2$ stored as a function of the pressure for an aquifer and two coal seams with different adsorption capacities. Reservoir size is the same for all three scenario, namely $100 \times 100 \times 2$ m. Coal density is 1400 kg/m$^3$

With particular reference to coal bed methane, the knowledge of its amount within the reservoir before starting its primary and enhanced recovery, i.e. the so-called Gas-In-Place (GIP), is important for several

## 7.3 The gas-in-place in coal seams

reasons. First, it is used for the design of the ECBM operation because, it defines both the reservoir pressure level where gas starts to be released during primary recovery and the amount of gas which can be recovered (Mares et al., 2009). This concept can be easily visualized with the help of Figure 7.8. The maximum holding capacity of a given coal seam is given by the sorption isotherm; comparison of the actual gas content with the value given by the sorption isotherm at the specific pressure and temperature condition, determines therefore the level of saturation of the coal seam. Saturated coals will desorb gas as soon as pressure is reduced, whereas for undersaturated coal seam, the so-called critical desorption pressure $P_C$ has to be reached (Bustin and Bustin, 2008). It is worth pointing out, that depending of the particular position on the isotherm, the amount of pressure drawdown required can be considerable, thus affecting the overall economics of the process (McElhiney et al., 1993). Finally, the amount of $CH_4$ which can be recovered is defined by the so-called abandonment pressure $P_A$, i.e. the bottom-hole pressure attainable for the well.

A reliable estimation of the GIP can be achieved through the so-called pressure coring method, which allows obtaining a coal sample at reservoir pressure, thus precluding any loss of gas (Diamond and Schatzel, 1998; Yee et al., 1993). The amount of gas actually contained in the sample can then be evaluated by letting the coal sample desorb $CH_4$ and by measuring its desorption rate and the overall amount released (Diamond and Schatzel, 1998). After having performed the desorption experiment, the same sample can be used for sorption analysis at the pressure and temperature conditions corresponding to the reservoir depth. Comparing the results of the two experiments allows defining the degree of gas saturation of the coal (Figure 7.8). It is commonly accepted, that coal seams usually generate more gas compared to the amount which can be

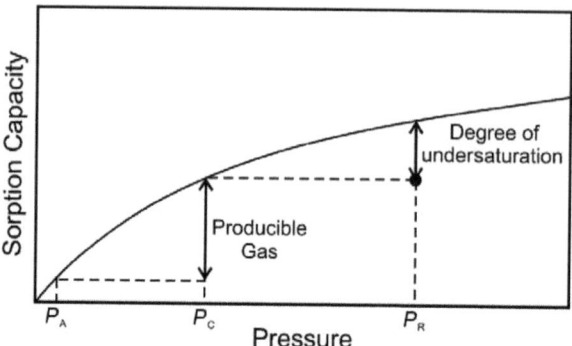

Figure 7.8: Actual degree of saturation of a coal seam as compared to the adsorption isotherm and determination of the pressures controlling the process: reservoir pressure $P_R$, critical desorption pressure $P_C$ and abandonment pressure $P_A$. From Bustin and Bustin (2008).

retained by the coal. Under this assumption, mainly two reasons can be found in the literature for explaining the undersaturation of coal seams (Hildenbrand et al., 2006; Bustin and Bustin, 2008): the gas leaked out of the reservoir, either as a gas or dissolved in water (Section 7.2), or the gas was lost during uplift and erosion. The former is an important information in the view of $CO_2$ storage in the same coal seam, since by indicating the amount of $CH_4$ that has left the coal bed, it provides an indirect information about the sealing efficiency of the cap rock and/or the hydrogeological regime of groundwater flow. Coal beds with a $CH_4$ content close to the maximum theoretical value given by the adsorption isotherm are therefore promising for a future ECBM project. The second aspect deals with the thermal history experienced by the coal seam during burial. Many coal basin have been uplifted from their maximum depth of burial; during this process, the reservoir temperature decreases and, being adsorption an exothermic process, this results in a increased sorption capacity. However, during uplift pressure decreases and the adsorption capacity reduces as well. Assuming that no additional gas has been generated after the time of maximum burial, in both cases the gas saturation of the coal seam changes.

## 7.4 Concluding remarks

In this chapter, the main mechanisms that prevent the $CO_2$ to leak out of the reservoir have been reviewed together with those controlling its release. Two flow behaviors have been identified, that can be seen as the two limiting cases of the real flow pattern: the diffusive and the convective flow. Both mechanisms have been studied independently with the help of one dimensional models in order to identify and quantify their time-scale. Even if the presented models assume a strong simplification of the real geological situation, they allow us to bring important conceptual insights concerning the leakage process from the reservoir. It has been shown that one of the most important characteristics of the cap rock is the size of the pores as it controls the capillary pressure. The convective flow is characterized by a sharp advancing saturation front, and as expected, it is the the most dangerous leaking process, since $CO_2$ moves as a gas towards the surface. The second part of this chapter focused on coal seams, and in particular on the interpretation of the gas-in-place (GIP), a quantity that allows to obtain information on the sealing efficiency of the cap rock overlying the coal seams.

## 7.5 Nomenclature

| | |
|---|---|
| $b$ | Langmuir constant [Pa$^{-1}$] |
| $D$ | Diffusion coefficient [m$^2$/s] |
| $D_p$ | Pore diffusivity [m$^2$/s] |
| $\varepsilon$ | Porosity [-] |
| $g$ | Gravitational acceleration [m/s$^2$] |
| $\theta$ | Contact angle between the gas/water separation surface [°] |
| $J$ | Molar flux [mol/m$^2$/s] |
| $k$ | Absolute permeability [mD] |
| $k_i$ | Permeability of phase $i$ [mD] |
| $k_{ri}$ | Relative permeability of phase $i$ [mD] |
| $\mu_i$ | Viscosity of phase $i$ [Pa.s] |
| $P_i$ | Pressure of phase $i$, [bar] |
| $P_s$ | Vapor pressure [bar] |
| $P_c$ | Capillary pressure [bar] |
| $r$ | Radius [r] |
| $\rho_i$ | Density of phase $i$ [mol/m$^3$] |
| $S$ | Gas saturation [-] |
| $\sigma_{gw}$ | gas/water interfacial tension [N/m] |
| $T$ | Temperature [°C] |
| $\tau$ | Tortuosity factor [-] |
| $u_i$ | Velocity of phase $i$ [m/s] |

**Subscripts and Superscripts**

| | |
|---|---|
| g | Gas phase |
| w | Water phase |
| 0 | Initial state |

## 7.5 Nomenclature

# 7. Containment in the reservoir

# Chapter 8

# Outlook

In this research project, several experimental and modeling tools have been developed allowing for a coal characterization aimed at assessing the potential of the given coal for an ECBM operation. This effort is motivated by the need of understanding the mechanisms acting during the process of gas displacement in coal beds. In particular, the issues which have been covered in this thesis are pure and competitive sorption experiments on coal of the gases involved in the process, being the former essential for the coal storage capacity estimates and the latter a prerequisite for the description of the displacement dynamics. The phenomenon of coal swelling and its effect on the coal permeability have been also investigated. The latter is in fact the parameter controlling the gas flow through the coal seam and affecting therefore the overall ECBM operation. All the outcomes of the above mentioned experimental studies are the needed information to be included in ECBM simulation studies, aimed at the description and design of ECBM processes. For this reason, a modeling work has been undertaken and the effects of the composition of the injected gas on the performance of the ECBM operations in terms of amount of $CO_2$ stored, $CH_4$ recovered and purity have been quantified. We believe that only a thorough understanding of the different mechanisms acting during ECBM achieved through the just mentioned experimental and theoretical studies will allow to critically assess the success or the failure of the performed field tests as well as the feasibility of future demonstration projects. Obviously, more research work and field tests under different geological settings are required to increase the confidence in ECBM as a possible approach for reducing carbon dioxide emissions. Some issues that need further studies are the long term stability and fate of $CO_2$ stored, the effect of impurities such as $SO_x$, $NO_x$ and $O_2$ on the process, the heterogeneity of the coal seam that makes the extrapolation of the lab results to the field challenging, and the effect of water on the storage capacity and displacement dynamics. Besides, in developing a new field test innovative methods have to be applied to characterize the 3D coal formation and to monitor properly the ECBM operation. In the following sections a brief outlook is given on the research which has been recently undertaken to investigate some of the issues just mentioned and that will be pursued in the future.

## 8.1 Gas sorption on wet samples

The gas sorption experiments presented in Chapter 2 and 3 have been carried out on dry samples. The assumption is often done that coal seams behave like dry reservoirs during ECBM recovery. In fact, already from the final stage of primary coal bed methane recovery, the amount of produced water becomes practically insignificant, as the coal bed has been dewatered to reduce the reservoir pressure (Durucan and Shi, 2009). However, depending on when the ECBM operation starts,

## 8.1 Gas sorption on wet samples

the coal seam may, at least partly, still be saturated with water. Laboratory experiments should therefore involve also measurements on wet samples, to better reproduce reservoir conditions. Of particular interest are sorption capacity estimates, as they affect the overall economics of the ECBM process, if the aim is to store as much $CO_2$ as possible. Unfortunately, techniques for the measurement of sorption isotherms on wet samples are not as well established as those on dry samples. All adsorption measurements show that wet coal uptakes $CO_2$ always less than dry coal, because of competitive water adsorption. However, the quantitative effect of moisture on $CO_2$ uptake is less certain because of the intrinsic difficulty of the measurements. It has been shown for instance that a coal containing about 3% moisture, corresponding to a relative humidity of 50%, adsorbs 30% less $CO_2$ than the corresponding dry one (Day et al., 2008c). Some have measured adsorption on coal samples as received (Siemons and Busch, 2007; Fitzgerald et al., 2005); others have prepared wet coal samples by equilibrating them in a sealed chamber with a saturated salt-solution of known water partial pressure (Day et al., 2008c; Krooss et al., 2002). In all cases one has to make the assumption that the adsorbed amount of water on coal remains constant as the gas adsorbs, i.e. that the experiment is carried out in such a way to prevent the gas from drying the sample. The fact that the water content of the coal sample may change for other reasons than competitive sorption, precludes any physically interpretation of the obtained results. This is believed to be one of the reasons why an inter-laboratory study using different techniques has shown unsatisfactory reproducibility of $CO_2$ adsorption data on wet coal, particularly above 80 bar (Goodman et al., 2007, 2004). In this context, the impact of several sources of error on the measured high-pressure adsorption isotherms have been recently discussed in detail (Sakurovs et al., 2008a). In our laboratory we have developed a technique to perform high-pressure gas sorption experiments on wet coal samples under a constant relative humidity atmosphere. In the next section the experimental procedure is presented and some preliminary results are shown.

### 8.1.1 Experimental procedure

As in the case of experiments on dry samples, prior to the sorption measurements, the coal samples is crushed and sieved to obtain the desired particle size. Subsequently, it is dried in an oven at 105°C under vacuum for 1 day. The moist sample was then prepared by equilibrating it over saturated salt solutions at the temperature of the experiments. The moisture content is then obtained by the difference in weight between moist and dry samples. Table 8.1 reports the water-salt solutions used for the experiments, together with the corresponding relative humidities, water vapor pressures and the obtained moisture level of the coal sample.

The same magnetic suspension balance used for the gas sorption experiments on dry coal samples (Chapter 2 and 3) is used to perform experiments on wet coal samples. Two main modifications are applied to the experimental set-up: first, all the parts which are not heated by the heating jacket of the magnetic suspension balance are carefully covered with a heating tape, that keeps the temperature at the same level as in the high-pressure cell to avoid water condensation. Secondly, a basket containing a saturated salt solution is placed on the bottom of the measuring cell, to control the relative humidity, as shown in Figure 8.1. The assumption is done, that, by ensuring a constant relative humidity inside the high-pressure cell, any drying of the sample during the experiments is avoided.

## 8.1 Gas sorption on wet samples

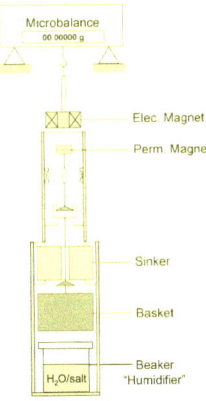

Figure 8.1: Magnetic Suspension Balance modified to perform gas sorption experiments at constant relative humidity.

Table 8.1: Salts used for the experiments, together with the corresponding obtained relative humidities, water vapor pressure and the moisture level of the coal sample.

| salt | NaCl | MgCl$_2$ | LiCl |
|---|---|---|---|
| **RH [%]** | 75 | 30 | 11 |
| $P^w$(45°C) [bar] | 0.072 | 0.029 | 0.011 |
| **moisture [%w.]** | 10.5 | 5.7 | 2.7 |

# 8. Outlook

A typical sorption experiment consists of the following steps: the high pressure cell containing the wet coal sample (about 3 g) and the beaker with the saturated salt solution is evacuated and the weight under vacuum is measured. Note that the lowest pressure which can be reached during the evacuation step is the vapor pressure given by the water/salt solution. Then, the system is filled with helium to obtain the volume of the metal parts and the wet coal sample ($V^0 + V^w$). After evacuating it again, the cell is filled with $CO_2$ and the weight is measured at the desired conditions, i.e.

$$M_1(\rho^b, T) = M_1^0 + m^w + m^t - \rho^b(V^0 + \Delta V^*) \tag{8.1}$$

where $M_1^0$ is the weight of the dry coal sample and metal parts under vacuum; $m^w$ and $m^t$ the amounts of water uptake by the coal and the amount adsorbed and absorbed of $CO_2$ on the coal respectively; $\Delta V^*$ is the contribution of the corresponding volumes, i.e. the volume increases given by gas and water sorption; $\rho^b$ is the mass density of the bulk phase and $V^0$ the sum of the volumes of the metal parts and the initial unswollen coal sample. As in the case of competitive sorption experiments (Chapter 3), the total excess sorption $m^{eas}$ can be directly obtained from the experiments and is given by

$$\begin{aligned} m^{eas}(\rho^b, T) &= M_1(\rho^b, T) - M_1^0 + \rho^b V^0 \\ &= m^* - \rho^b \Delta V^* \end{aligned} \tag{8.2}$$

where $m^* = m^w + m^t$ is the total uptake (water and $CO_2$) adsorbed on and absorbed in the coal. In order to obtain the excess sorption of $CO_2$, the assumption is done that $m^w$ is constant during the experiments and

## 8.1 Gas sorption on wet samples

that the composition of the bulk phase corresponds practically to pure $CO_2$, i.e.,

$$\begin{aligned}
m_{CO_2}^{eas}(\rho^b, T) &= M_1(\rho^b, T) - (M_1^0 + m^w) + \rho^b V^0 \\
&= m^t - \rho^b \Delta V^*
\end{aligned} \quad (8.3)$$

Note that only the mass of adsorbed water is assumed to be constant, and not its volume. As in the case of gas mixture, no distinction is made between the different components and only one adsorbed phase volume (of the mixture) is considered (Chapter 3). Moreover, it is worth pointing out that $CO_2$ may dissolve into the adsorbed water, thus providing additional storage capacity. However, as long as one is interested in quantifying the total $CO_2$ uptake, this is not a problem since it is taken into account in the weight measurement of the balance.

### 8.1.2 Results and discussion

In Figure 8.2 are shown the results of the sorption experiments carried out at 45°C with a wet coal sample (coal sample I2, from the Sulcis Coal Province). In particular the total sorption, as obtained from Eq.(8.2) (filled symbols), and the $CO_2$ excess sorption (empty symbols), as obtained with Eq.(8.3) are plotted as a function of the bulk density. The dashed line corresponds to the excess sorption of pure $CO_2$ on the same dried coal sample. It can be seen that the total uptake ($H_2O + CO_2$) increases with increasing relative humidity, as a consequence of an increased amount of water on the sample. On the other hand, the $CO_2$ excess sorption decreases with increasing relative humidity. This effect can be attributed to the competitive sorption mechanism between water

# 8. Outlook

and $CO_2$. It should be pointed out that during the sorption experiments, at different intervals the pressure in the measuring cell was completely released and the weight of the coal sample measured. Comparison with the initial weight of moist sample confirmed a constant moisture content in the range of $\pm\ 0.1\%$.

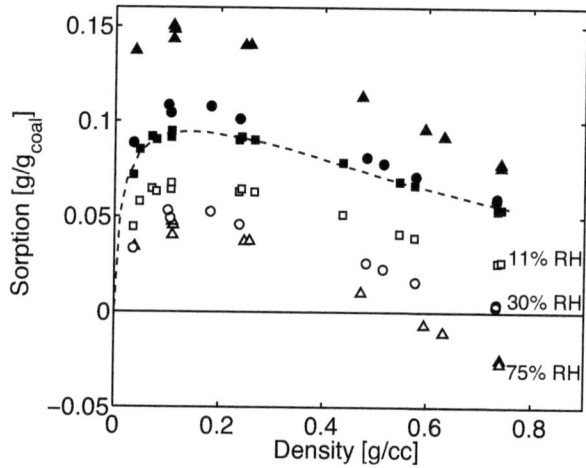

Figure 8.2: Total (filled symbols) and $CO_2$ (empty symbols) excess sorption on a wet coal (sample I2) as a function of the bulk density at 45°C. The dashed line corresponds to the pure $CO_2$ dry case, whereas points are experimental results: ($\triangle$, $\blacktriangle$), 75%, ($\circ$, $\bullet$) 30% and ($\square$, $\blacksquare$) 11% relative humidity.

These preliminary results suggest that water content affect considerably the $CO_2$ sorption capacity, in agreement with other studies (Day et al., 2008c). Further experiments are however needed to better quantify the effect of moisture on the sorption capacity of $CO_2$. The interaction of water with coal is not trivial as water can be present in a number of physical states on the coal, and this situation is further complicated when a third component (in this case $CO_2$) is added to the system. The main concern

## 8.1 Gas sorption on wet samples

is to what extent competitive water-$CO_2$ sorption can be regarded as the "usual" competitive sorption between two gases (see Chapter 3). The issues that need further clarification are listed as follows:

- Liquid water is present in the beaker placed in the measuring cell. The extent to which the partial mutual solubility between water and $CO_2$ is affecting the fugacity of the two components, and as consequence the thermodynamic equilibrium of adsorption has to be quantified.

- Pure water sorption isotherm can be measured up to the water saturation pressure at the given temperature of the experiment. To understand the effect of a higher pressure on the water content of the sample, an experiment could be carried out using the non-adsorbing helium, allowing therefore to directly estimate the amount of water adsorbed.

- In the experiments described above, the maximum vapor pressure reached was well below the saturation water pressure. It should be tested if this is enough to avoid water condensation on the metal parts (basket and sinker) in the measuring cell.

- Water and $CO_2$ isotherms are of different type: the former is of Type II (Day et al., 2008c), whereas the latter is of Type I. A modeling attempt should be made (tentative mixture isotherm), to describe the competition between the two components. Moreover, the mechanism of water sorption will rather be pore filling, than monolayer coverage.

- Once a protocol has been defined for the experiments with $CO_2$, the method could be applied to other gases ($CH_4$ and $N_2$) and should be extended to mixtures of these gases.

# 8. Outlook

## 8.2 Displacement experiments

In Chapter 5, an experimental protocol has been developed and results presented of pure gas injection experiments into coal cores confined by an external pressure. These experiments were intended to improve the knowledge on the different mechanisms acting during $CO_2$ storage in coal seams and in particular on those related to permeability. In order to make a step further in the understanding of the displacement mechanisms during ECBM, the technique has to be extended to mixture, allowing to perform small-scale ECBM experiments in the laboratory. In the following is described the concept developed to carry out these experiments.

The transient step method, which has been shown to be very effective for pure gases, can be applied to the experiments with gas mixture as well and the experimental procedure would therefore be very similar.

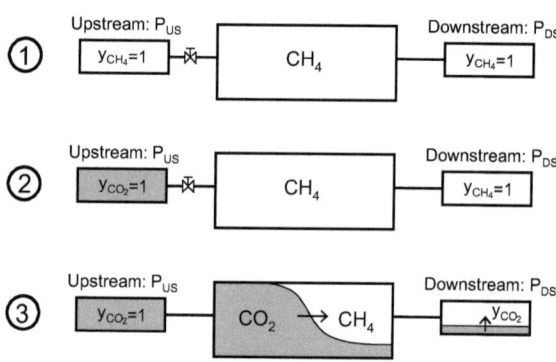

Figure 8.3: Schematic for the procedure adopted for the flow experiment involving gas mixtures.

As shown in Figure 8.3 with a simplified schematic, once the system

## 8.2 Displacement experiments

upstream reservoir-coal core-downstream reservoir has been let equilibrate at a given pressure with methane (1), the valve connecting the upstream reservoir is closed, and the upstream reservoir is filled with the gas to be injected ($CO_2$) at a higher pressure (2). The valve is the opened (3) and the system is let to equilibrate at a new pressure level. Beside the pressure reading during the transient step, an additional information is needed to completely characterize these experiments, namely the composition of the fluid phase in the downstream reservoir, which changes due to the $CO_2/CH_4$ displacement. During the course of the experiment therefore, sample from the downstream reservoir are taken and their composition analyzed by a gas chromatograph.

Figure 8.4 shows a detail of the downstream reservoir, with the sampling loop to be sent to the gas chromatograph. Two considerations are worth making with respect to this point. First, when taking a sample, it is crucial that it has the same composition as the entire reservoir. A good mixing can be ensured, by using a reservoir with a relatively large diameter (in our case about 2 cm), therefore avoiding pressure drops. Secondly, the sample volume has to be large enough for a reliable analysis, yet small enough to have minimal impact on the system being studied. Preliminary experiments have shown that extracting a sample of 7 ml (atmospheric pressure) results in a pressure drop of about 0.3 to 0.4 bar in the downstream reservoir.

As in the case of pure gases, a model is needed to describe the experiments allowing to obtain information on the permeability changes of the coal sample during the injection experiment. Figure 8.5 shows simulation results of an $CO_2/CH_4$ experiment, where $CO_2$ is injected at 40 bar into a coal sample, that has been pre-saturated with $CH_4$ at 20 bar. Confining pressure is 100 bar. In Figure 8.5a the pressure transients are plotted, whereas Figure 8.5b shows the change

8. Outlook

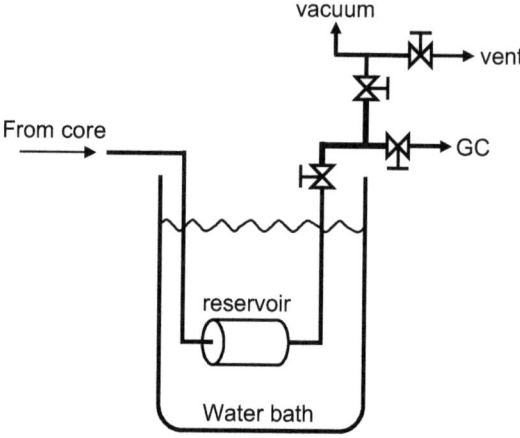

Figure 8.4: Detail of the downstream reservoir, with the sampling loop (thicker lines) to be sent to the gas chromatograph. Each time before taking a new sample, the sampling loop is evacuated.

in composition in the downstream reservoir. As expected, reservoir equilibrates at a new pressure level and the molar fraction of $CO_2$ in the downstream reservoir increases with time reaching a plateau once the pressure gradient becomes negligible. A first important conclusion can be drawn from these results. In order to get some insights on the adsorption/desorption displacement, gas flow through the coal has to be slow enough to allow reaching adsorption/desorption equilibrium between the gases. From the pure gas experiments, the adsorption time constant obtained for each gas was in the order of few days, in agreement with other studies. As a consequence if gas flow is too fast, the measured concentration change in the downstream reservoir simply corresponds to a gas exchange in the fluid phase, without any effect of gas sorption. To avoid this the confining pressure will be set at a level

## 8.2 Displacement experiments

which allows to obtain the desired experiment speed.

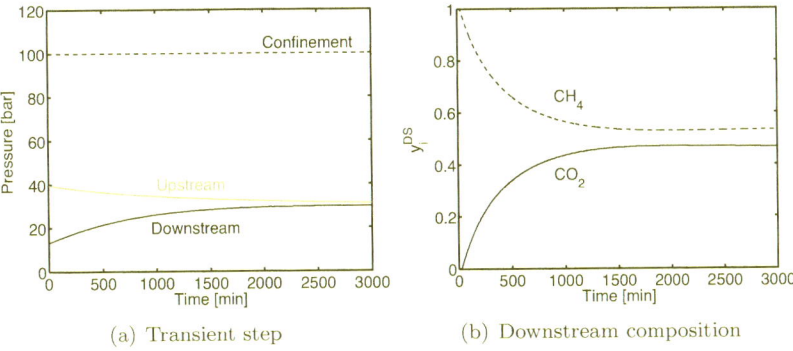

(a) Transient step

(b) Downstream composition

Figure 8.5: Simulation results of an $CO_2/CH_4$ experiment, where $CO_2$ is injected at 40 bar into a coal sample, that has been pre-saturated with $CH_4$ at 20 bar. Confining pressure is 100 bar. Pressure transients in both reservoir (a) and compositions changes in downstream reservoir (b) as a function of time.

Finally, it was shown in Chapter 6, that injection of flue gas into a coal bed for ECBM recovery is a promising alternative to pure $CO_2$ injection to avoid the injectivity problems encountered in the field tests. It is therefore suggested, that the experimental activity will focus not only on the $CO_2/CH_4$ system, but on $N_2/CH_4$ as well as with mixture of $CO_2$ and $N_2$ displacing methane.

# 8. Outlook

# Appendix A

# Proximate analysis of coal

## A.1 Thermogravimetric analysis (TGA)

Coalification is the process by which plant material has been progressively altered through peat, lignite, subbituminous, and bituminous coals to anthracite (Van Krevelen, 1981). During this time, the coal is exposed to a temperature gradient as it is buried deep in the earth. Since this process is mainly controlled by time and temperature, thermal analysis of coal should be able to show measurable differences, corresponding to the stage of the coalification reached by the specific sample (Huang et al., 1999). Exactly because of this reasons, thermal methods of analysis are widely applied in the coal industry.

Coal can be roughly characterized by volatile and nonvolatile components. Volatile fraction can either be organic such as methane and other low molecular weight compounds or inorganic such as moisture. Non-volatile compounds can also be organic (maceral components) and inorganic (mineral matter).

Proximate analysis is the name given to the standard method for measurement of the constituent of coals and cokes, as prescribed by national standards organizations (Ottaway, 1982). The outcome of this analysis is an approximate estimation of moisture, volatile matter (VM), fixed carbon (FC) and ash content of a coal sample. Following the definition given by Mukhopadhyay (1993), the terms fixed carbon, volatile matter and ash are related to the experimental conditions for the residue (ash) or liberated material after combustion. Fixed carbon represent the residue (other than ash) after the moisture and volatile matter are liberated from the coal. Mineral matter is the major contributor to ash yield.

Quantification of each of the four components of coal by standards methods requires the samples to be heated under specified conditions at a selected temperature. An alternative method to provide a measure of the proximate analysis of coal is given by thermogravimetry (Ottaway, 1982). The measurement is completely automatized and is therefore much less time consuming than the conventional method used. Because of the rapid improvement and automation of TGA equipments, thermograms of coal are now comparatively easy and precise measurements to make (Huang et al., 1999).

## A.2 Experimental procedure

The experiments were carried out in a thermogravimetric system (TGA, Netzsch STA 409 CD). An approximately 10 mg sample of each coal was used for the experiments. The weight-loss curve is recorded during the

## A.2 Experimental procedure

measurements and can be used to obtain the proximate composition of the coal sample. The steps of the procedure adopted in this work are summarized as follows (Huang et al., 1999):

- Heat approximately 10 mg of the sample at 10°C/min from ambient temperature to 150°C in 25 mL/min of $N_2$.

- Hold for 10 min at 150°C in 25 mL/min of $N_2$. The moisture content is determined by the weight change.

- Continue to heat the sample at 10°C/min to 950°C in 25 mL/min of $N_2$.

- Hold for 7 min in the presence of $N_2$ at 25 mL/min. The volatile matter (VM) is determined by the weight change.

- Switch to oxygen (25 mL/min) at 950°C and hold for 30 min. The weight change measures the fixed carbon (FC), and the weight of the residue gives the ash content.

A sample TGA thermogram for coal sample J1 is shown in Figure A.1, where the weight-loss curve is reported together with the temperature profile in the cell. The three stages of weight loss are shown and the component loss which relates to each of these is indicated. The same procedure has been adopted for all the coal samples investigated in this study. Results of proximate analysis are reported in Chapter 2 (Table 2.1).

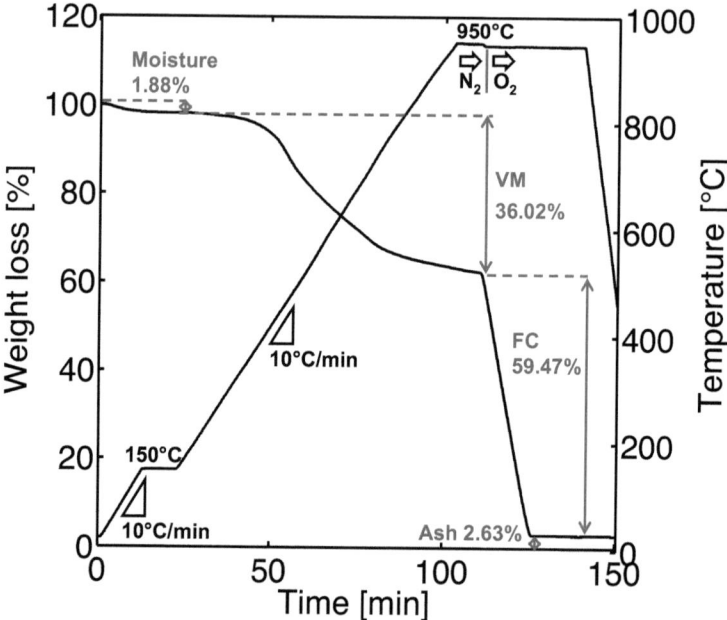

Figure A.1: Thermogravimetric curve for a coal using the programme described for coal sample J1. Weight losses and isothermal temperatures are indicated.

# Appendix B

# List of Figures

1.1 Schematic of an ECBM operation, where captured $CO_2$ from a power plant is injected into the coal seam and $CH_4$ is produced. Injection and production wells can in general be more than one. .................... 6

1.2 $CO_2$ density, pressure and temperature as a function of depth to be expected for underground $CO_2$ storage, assuming a geothermal gradient of 25°C/km from 15°C at the surface, and specific weight of soil and water of 22.62 kN/m$^3$ and 10.18 kN/m$^3$, respectively. $CO_2$ density increases rapidly at 800 m depth, when $CO_2$ reaches its supercritical state. Generally, the hydrostatic pressure is taken as the criterion to determine $CO_2$ injection pressure. The lithostatic pressure is the pressure exerted on the coal bed by the surrounding rock (also called geostatic pressure). ..................... 8

## B. List of Figures

2.1 Reflected light microphoto of a polished section of coal sample I2. Scale bar 200 $\mu$m. ................ 19

2.2 $CO_2$ excess sorption amount $n^{eas}$ (●) obtained for coal sample I2 at 45°C as a function of the bulk density. The total uptake $n^t = n^a + n^s$ (○) has been obtained by applying the graphical estimate method (Sudibandriyo et al., 2003b). ................................ 22

2.3 $CO_2$ sorption isotherm $n^t$ (○) obtained for coal sample I2 at 45°C as a function of the bulk density. Lines represent model results from two different isotherm equations: Bi-Langmuir model (black solid line) with corresponding component contributions (dashed lines), and Langmuir-like model, Eq.(2.7) (gray solid line). ............ 28

2.4 $CO_2$ excess sorption isotherm $n^{eas}$ (●) obtained for coal sample I2 at 45°C as a function of the bulk density and corresponding total uptake $n^t$ (○). Lines represent model results from two different isotherm equations: Langmuir-like model, Eq.(2.9) (black lines) and DR equation combined with Henry's law, Eq.(2.8) (gray lines). ........ 30

2.5 Comparison among different laboratories. $CO_2$, $CH_4$ and $N_2$ molar excess sorption $n^{eas}$ on coal samples (a) A1 and (b) A2 at 55°C as a function of the bulk density measured at CSIRO (Newcastle, Australia) (open symbols) (Sakurovs et al., 2007) and in our lab (closed symbols). . 34

2.6 High pressure pure sorption isotherms on coal. $CO_2$, $CH_4$ and $N_2$ molar excess sorption $n^{eas}$ as a function of the bulk density $\rho^b$ on eight coal samples measured at 45°C. Symbols are experimental points, whereas lines are fitted Langmuir curves. Symbols: I1 ($\Diamond$), I2 ($\blacklozenge$), J1 ($\square$), A1 ($\blacksquare$), A2 ($\circ$), A3 ($\bullet$), S2 ($\triangle$), S3 ($\blacktriangle$). . . . . . . . . . . . . 38

2.7 Total $CO_2$ uptake $n^t$ as a function of the bulk density $\rho^b$ on eight coal samples measured at 45°C. Symbols are experimental points, whereas lines are fitted Langmuir curves. Symbols: I1 ($\Diamond$), I2 ($\blacklozenge$), J1 ($\square$), A1 ($\blacksquare$), A2 ($\circ$), A3 ($\bullet$), S2 ($\triangle$), S3 ($\blacktriangle$). . . . . . . . . . . . . . . . . 39

2.8 $CO_2$, $CH_4$ and $N_2$ molar excess sorption $n^{eas}$ on coal sample I1 measured at three different temperatures, namely 33 ($\triangle$), 45 ($\circ$) and 60°C ($\square$). Symbols are experimental points, whereas lines are model results: fitted excess Langmuir curves (dashed lines) and their corresponding absolute isotherms (solid lines). . . . . . . . . . . . . . . . 43

2.9 $CO_2$ molar excess sorption $n^{eas}$ as a function of the bulk density $\rho^b$ measured at 45°C on coal samples collected at different depths: S1 ($\triangle$) at 1743 m, S2 ($\square$) at 1586 m and S3 ($\circ$) at 1701 m. Symbols are experimental points, whereas lines are model results: fitted excess Langmuir curves (dashed lines) and their corresponding absolute isotherms (solid lines). . . . . . . . . . . . . . . . . . . . . 45

2.10 Maximum sorption capacity $n^{max}$ of $CO_2$ ($\circ$), $CH_4$ ($\triangle$) and $N_2$ ($\square$) at 45°C as a function of vitrinite reflectance $R_o$. Symbols are experimental points, whereas lines are fitted parabolic curves. . . . . . . . . . . . . . . . . . . . . . 47

# B. List of Figures

3.1 Scheme of the setup for competitive adsorption measurements at up to 300 bar and 80°C. For better visibility, the calibrated void volume of the system is connected by thick solid lines. . . . . . . . . . . . . . . . . . . . . . . . . . 60

3.2 Langmuir sorption isotherms at 45°C as a function of pressure for $CO_2$ (○), $CH_4$ (□) and $N_2$ (△) for a coal from the Sulcis coal province. Symbols are experimental points, whereas lines are fitted Langmuir curves. . . . . . . . . . . . 64

3.3 Binary high-pressure sorption isotherms of $CO_2$ (1) and $N_2$ (2) on the Sulcis coal at 45°C at 40 (a), 100 (b) and 160 bar (c) as a function of gas composition $y_1$ ($CO_2$). Symbols are experimental points, whereas lines are the predicted extended Langmuir curves. Symbols: $CO_2$ (△), $N_2$ (□), total (●). . . . . . . . . . . . . . . . . . . . . . . 68

3.4 x-y diagram for $CO_2/N_2$ mixture at 45°C at three different pressures 40 (○), 100 (△) and 160 bar (□). Symbols are experimental points, whereas lines are extended Langmuir predicted curves. . . . . . . . . . . . . . . . . . . . . 69

3.5 Binary sorption isotherms of $CO_2$ (1) and $CH_4$ (2) on the Sulcis coal at 45°C at 40 (a), 100 (b) and 160 bar (c) as a function of gas composition $y_1$ ($CO_2$). Symbols are experimental points, whereas lines are the predicted extended Langmuir curves. Symbols: $CO_2$ (△), $CH_4$ (□), total (●). . . . . . . . . . . . . . . . . . . . . . . . . . . . . . . 71

3.6 x-y diagram for $CO_2/CH_4$ mixture at 45°C at three different pressures 40 (○), 100 (△) and 160 bar (□). Symbols are experimental points, whereas lines are extended Langmuir predicted curves. . . . . . . . . . . . . . . . . . . . . 72

3.7 Binary high-pressure sorption isotherms of $CH_4$ (1) and $N_2$ (2) on the Sulcis coal at 45°C at 40 (a), 100 (b) and 160 bar (c) as a function of gas composition $y_1$ ($CH_4$). Symbols are experimental points, whereas lines are the predicted extended Langmuir curves. Symbols: $CH_4$ (△), $N_2$ (□), total (●). . . . . . . . . . . . . . . . . . . . . . . . 74

3.8 x-y diagram for $CH_4/N_2$ mixture at 45°C at three different pressures 40 (○), 100 (△) and 160 bar (□). Symbols are experimental points, whereas lines are extended Langmuir predicted curves. . . . . . . . . . . . . . . . . . . . . . 75

3.9 Sorption amount $n_i^t$ of component $i$ per unit mass of coal at 45°C as a function of the total pressure, $P$. Feed composition of ternary mixture: 33.3% $CO_2$, 33.3% $CH_4$, 33.4% $N_2$. Symbols: experimental data; Lines: extended Langmuir equation. . . . . . . . . . . . . . . . . . . . . . . 77

4.1 Schematic of the high-pressure view cell used for the swelling experiments (left) and obtained image of the coal sample (right). . . . . . . . . . . . . . . . . . . . . . . 85

4.2 Swelling of an unconstrained dry disc (Sulcis coal) as a function of the time when exposed to $CO_2$ at 45°C. Swelling is assumed to be isotropic. . . . . . . . . . . . . . . 88

4.3 Swelling of an unconstrained dry disc (Ribolla coal), as a function of the pressure, $P$, of $CO_2$, $CH_4$ and $N_2$ at 45°C. Swelling is assumed to be isotropic. . . . . . . . . . . . . . . 89

4.4 Swelling of an unconstrained dry disc (Sulcis Coal) as a function of the pressure, $P$, of $CO_2$, $CH_4$, $N_2$ and He at 45°C. Swelling is assumed to be isotropic. . . . . . . . . . 90

B. List of Figures

4.5 Dimensionless volumetric swelling $s/s^{max}$ for coal samples (a) I1 and (b) I2 at 45°C as a function of dimensionless adsorbed and absorbed amount $n^t/n^{max}$ for $CO_2$, $CH_4$ and $N_2$ . . . . . . . . . . . . . . . . . . . . . . . . . . . 94

5.1 (a) Adsorption and (b) swelling isotherms at 45°C as a function of pressure for $CO_2$ (○) and $N_2$ (□) measured on a coal sample from the Sulcis coal province. Solid lines correspond to the Langmuir model. . . . . . . . . . . . . . 103

5.2 Setup used in this study for the permeability measurements under confined conditions. . . . . . . . . . . . . . 106

5.3 Example of an experimental transient step: confining pressure $P_c$ (△), upstream $P_{US}$ (○) and downstream $P_{DS}$ (□) reservoir pressures as a function of time. . . . . . . . 116

5.4 Time to complete 50% of the imposed transient step, $\tau_{0.5}$ (logarithmic scale) as a function of the effective pressure $P_e = P_c - P_{eq}$ on the sample when Helium (○), $N_2$ (▲) and $CO_2$ (□) were injected. Dashed lines represent model results. . . . . . . . . . . . . . . . . . . . . . . . . . . . . 117

5.5 Transient steps measurements at 45°C when Helium is injected: (a) confining pressure kept constant and (b) varying confining pressure. Confining pressure $P_c$ (△), upstream $P_{US}$ (○) and downstream $P_{DS}$ (□) reservoir pressures as a function of time. Solid lines correspond to model results. . . . . . . . . . . . . . . . . . . . . . . . . . 120

5.6 Model predicted coal sample relative porosity and permeability as a function of the effective pressure $P_e$ for Helium at 45°C with confining pressure $P_c$ kept constant at ($\triangle$) 60, ($\circ$) 100 and ($\square$) 140 bar, respectively. Lines are model results and symbols represent the corresponding permeability and porosity obtained at the end of each transient step. . . . . . . . . . . . . . . . . . . . . . . . . . . . 121

5.7 Transient steps measured at 45°C when $CO_2$ is injected: (a) confining pressure kept constant and (b) varying confining pressure. Confining pressure $P_c$ ($\triangle$), upstream $P_{US}$ ($\circ$) and downstream $P_{DS}$ ($\square$) reservoir pressures as a function of time. Solid lines correspond to model results. . . . 124

5.8 Transient steps measured at 45°C when $N_2$ is injected by keeping the confining pressure constant at 100 bar. Confining pressure $P_c$ ($\triangle$), upstream $P_{US}$ ($\circ$) and downstream $P_{DS}$ ($\square$) reservoir pressures as a function of time. Solid lines correspond to model results. . . . . . . . . . . . . . . . 125

5.9 Model predicted relative porosity and permeability as a function of the equilibrium gas pressure $P_{eq}$ of (a) $CO_2$ and (b) $N_2$ with confining pressure kept constant at 100 bar when only the effective pressure term (dotted line), the swelling contribution (dashed line) and both terms (solid line) are taken into account. Lines are model results; symbols are the corresponding permeability and porosity obtained at the end of each transient step: Helium ($\triangle$) and $CO_2$ or $N_2$ ($\circ$). . . . . . . . . . . . . . . . . 126

B. List of Figures

5.10 Transient steps measured at 45°C when (a) He and (b) $CO_2$ are injected in the closed hydrostatic cell, i.e. without controlling the confining pressure. Confining pressure $P_c$ ($\triangle$), upstream $P_{US}$ ($\circ$) and downstream $P_{DS}$ ($\square$) reservoir pressures as a function of time. Solid lines correspond to model results. . . . . . . . . . . . . . . . . . . . . . . . 131

6.1 Langmuir sorption (a) and swelling (b) isotherms at 45°C as a function of pressure for $CO_2$ (solid line), $CH_4$ (dashed line) and $N_2$ (dotted line) for a coal from the Sulcis coal province. . . . . . . . . . . . . . . . . . . . . . . . . . . . . . 144

6.2 Model predicted relative permeability as a function of the $CO_2$ gas pressure $P$ with confining pressure kept constant at 100 bar when only the effective pressure term (dotted line), the swelling contribution (dashed line) and both terms (solid line) are taken into account ($C_1$=225.7 GPa$^{-1}$ and $C_2$=134.4). . . . . . . . . . . . . 147

6.3 Permeability ratio $k/k_0$ as a function of pressure $P$ under different injection scenarios (solid lines, Pure $CO_2$, 80:20/$CO_2$:$N_2$, 50:50/$CO_2$:$N_2$, 20:80/$CO_2$:$N_2$, pure $N_2$) for Case A (a) and Case B (b). The dashed line corresponds to the primary recovery scenario (pure $CH_4$). . . . . 154

6.4 Concentration profiles of $CO_2$, $CH_4$ and $N_2$ along the coal bed axis for three different injection scenarios: pure $CO_2$ (a) pure $N_2$ (b) and 50:50/$CO_2$:$N_2$ (c). . . . . . . . . . . . 156

6.5 Flow rates of $CO_2$, $CH_4$ and $N_2$ at the production well as a function of time for three different injection scenarios: pure $CO_2$ (a) pure $N_2$ (b) and 50:50/$CO_2$:$N_2$ (c). . . . . . 157

6.6 $CH_4$ purity (a) and $CH_4$ recovery (b) as a function of time for different ECBM schemes with different injection compositions (Pure $CO_2$, 80:20/$CO_2$:$N_2$, 50:50/$CO_2$:$N_2$, 20:80/$CO_2$:$N_2$, pure $N_2$). . . . . . . . . . . . . . . . . . . . . . 159

6.7 $CH_4$ purity (a) and $CH_4$ recovery (b) as a function of the amount of $CO_2$ injected for different ECBM schemes with different injection compositions (Pure $CO_2$, 80:20/$CO_2$:$N_2$, 50:50/$CO_2$:$N_2$, 20:80/$CO_2$:$N_2$). . . . . . . 160

6.8 Amount of $CO_2$ injected (a) and stored (b) as a function of time for different ECBM schemes with different injection compositions (Pure $CO_2$, 80:20/$CO_2$:$N_2$, 50:50/$CO_2$:$N_2$, 20:80/$CO_2$:$N_2$). . . . . . . . . . . . . . . . . . . . . . . . . 162

6.9 $CH_4$ purity (a) and $CH_4$ recovery (b) as a function of time for different ECBM schemes with different injection compositions (Pure $CO_2$, 80:20/$CO_2$:$N_2$, 50:50/$CO_2$:$N_2$, 20:80/$CO_2$:$N_2$, pure $N_2$). . . . . . . . . . . . . . . . . . . . . . 163

6.10 Amount of $CO_2$ injected (a) and stored (b) as a function of time for different ECBM schemes with different injection compositions (Pure $CO_2$, 80:20/$CO_2$:$N_2$, 50:50/$CO_2$:$N_2$, 20:80/$CO_2$:$N_2$, pure $N_2$). . . . . . . . . . . . . . . . . . . . . . 165

7.1 Schematic of the three main trapping configurations: structural traps, i.e. folds (A) and faults (B) and stratigraphic traps (C). . . . . . . . . . . . . . . . . . . . . . 171

# B. List of Figures

7.2 Trapping by capillary phenomena. (a) Schematic of capillary sealing mechanism in a pore throat of a seal rock. Modified from Li et al. (2005). (b) Capillary pressure curves calculated with the Young-Laplace equations for a $CO_2$-water system, with wetting angle $\theta = 45°$ and interfacial tension $\sigma_{nw}$ ranging from 0.045 and 0.078 N/m, as reported by Hildenbrand et al. (2004). . . . . . . . . . . . . 172

7.3 $CO_2$ concentration profiles in the 1 m thick caprock at different time intervals. Diffusion coefficient: $D=1\times10^{-9}$ $m^2/s$. 181

7.4 $CO_2$ cumulative flux obtained at the upper caprock boundary, i.e. at $z=1$ m, for three different diffusion coefficients. . . . . . . . . . . . . . . . . . . . . . . . . . . 182

7.5 Gas saturation profiles in the 1 m thick cap rock at different time intervals. Cap rock permeability $k=1\times10^{-6}$ mD. 183

7.6 $CO_2$ cumulative flux obtained at the upper cap rock boundary, i.e. at $z=1$ m, for four different permeability. Diffusive flux is shown for the sake of comparison (dashed line). . . . . . . . . . . . . . . . . . . . . . . . . . . . . . . 184

7.7 Amount of $CO_2$ stored as a function of the pressure for an aquifer and two coal seams with different adsorption capacities. Reservoir size is the same for all three scenario, namely $100 \times 100 \times 2$ m. Coal density is 1400 $kg/m^3$ . . 186

7.8 Actual degree of saturation of a coal seam as compared to the adsorption isotherm and determination of the pressures controlling the process: reservoir pressure $P_R$, critical desorption pressure $P_C$ and abandonment pressure $P_A$. From Bustin and Bustin (2008). . . . . . . . . . . . . . . . 188

8.1  Magnetic Suspension Balance modified to perform gas sorption experiments at constant relative humidity. . . . . 197

8.2  Total (filled symbols) and $CO_2$ (empty symbols) excess sorption on a wet coal (sample I2) as a function of the bulk density at 45°C. The dashed line corresponds to the pure $CO_2$ dry case, whereas points are experimental results: ($\triangle$, $\blacktriangle$), 75%, ($\circ$, $\bullet$) 30% and ($\square$, $\blacksquare$) 11% relative humidity.200

8.3  Schematic for the procedure adopted for the flow experiment involving gas mixtures. . . . . . . . . . . . . . . . . 202

8.4  Detail of the downstream reservoir, with the sampling loop (thicker lines) to be sent to the gas chromatograph. Each time before taking a new sample, the sampling loop is evacuated. . . . . . . . . . . . . . . . . . . . . . . . . . 204

8.5  Simulation results of an $CO_2/CH_4$ experiment, where $CO_2$ is injected at 40 bar into a coal sample, that has been pre-saturated with $CH_4$ at 20 bar. Confining pressure is 100 bar. Pressure transients in both reservoir (a) and compositions changes in downstream reservoir (b) as a function of time. . . . . . . . . . . . . . . . . . . . . . . 205

A.1  Thermogravimetric curve for a coal using the programme described for coal sample J1. Weight losses and isothermal temperatures are indicated. . . . . . . . . . . . . . . . . . 210

# Appendix C

# List of Tables

1.1 ECBM field tests. Well configuration: sw, single-well; 2w, two-well; mw, multi-well. . . . . . . . . . . . . . . . . . . . 11

2.1 Main properties of the nine coals investigated. [a]Ref. (Nagra, 1989). [b]Ref. (Sakurovs et al., 2007). . . . . . . . . . . . 20

2.2 Model parameters for $CO_2$ sorption on coal sample I2 at 45°C from two different isotherm equations: Bi-Langmuir model and Langmuir-like model, Eq.(2.7). . . . . . . . . . . 29

2.3 Model parameters for $CO_2$ sorption on coal sample I2 at 45°C from two different isotherm equations: Langmuir-like model, Eq.(2.9) and DR equation combined with Henry's law, Eq.(2.8). . . . . . . . . . . . . . . . . . . . . . 31

2.4 Langmuir model parameters for $CO_2$, $CH_4$ and $N_2$ sorption on coal samples A1 and A2 at 55°C. . . . . . . . . . . 35

2.5 Langmuir model parameters for $CO_2$, $CH_4$ and $N_2$ sorption on different coal samples at 45°C. . . . . . . . . . . . . 37

2.6 Langmuir model parameters for $CO_2$, $CH_4$ and $N_2$ sorption on coal samples I1 at three temperatures (33, 45 and 60°C). . . . . . . . . . . . . . . . . . . . . . . . . . . . . . . 42

3.1 Single component Langmuir equation parameters for $CO_2$, $CH_4$ and $N_2$ sorption on the coal sample used in this study at 45°C. . . . . . . . . . . . . . . . . . . . . . . . 64

3.2 Binary $CO_2$ (1) and $N_2$ (2) sorption data on the coal sample used in this study at 45°C and at 40, 100 and 160 bar. . . . . . . . . . . . . . . . . . . . . . . . . . . . . . . 67

3.3 Binary $CO_2$ (1) and $CH_4$ (2) sorption data on the coal sample used in this study at 45°C and at 40, 100 and 160 bar. . . . . . . . . . . . . . . . . . . . . . . . . . . . . . . 70

3.4 Binary $CH_4$ (1) and $N_2$ (2) sorption data on the coal sample used in this study at 45°C and at 40, 100 and 160 bar. 73

3.5 Ternary $CH_4$ (1) and $N_2$ (2) and $N_2$ (3) sorption data on the coal sample used in this study at 45°C. The gas phase composition was determined by gas chromatography. . . . 76

4.1 Studies reporting swelling measurements on coal. Maximum pressure, $P_{max}$ is given in bar. . . . . . . . . . . . 83

4.2 Properties of the two Italian coal samples investigated. . . 84

4.3 Experimental swelling data of pure $CO_2$, $CH_4$, $N_2$ on Ribolla coal (I1). . . . . . . . . . . . . . . . . . . . . . . . . 88

4.4 Experimental swelling data of pure $CO_2$, $CH_4$, $N_2$ and He on a coal disc from the Sulcis Coal Province (I2). . . . . . 89

| | | |
|---|---|---|
| 4.5 | Langmuir model parameters for $CO_2$, $CH_4$ and $N_2$ adsorption and swelling on coal samples I1 and I2 at 45°C. | 92 |
| 5.1 | Model input parameters | 102 |
| 5.2 | Langmuir isotherm parameters | 104 |
| 5.3 | Estimated values of the model fitting parameters | 118 |
| 5.4 | Porosity and permeability data at 45°C obtained at the end of each transient step when Helium is injected. | 122 |
| 5.5 | Porosity and permeability data at 45°C obtained at the end of each transient step when $CO_2$ is injected. | 128 |
| 5.6 | Porosity and permeability data at 45°C obtained at the end of each transient step when $N_2$ is injected. | 128 |
| 6.1 | Langmuir constants for the sorption and swelling isotherms for the coal from the Sulcis Coal Province. | 143 |
| 6.2 | Thermodynamic properties of $CO_2$, $CH_4$ and $N_2$ for the Peng-Robinson EOS. | 143 |
| 6.3 | Constants $C_1$ ($Pa^{-1}$) and $C_2$ of Eq.(6.8) as obtained from different permeability models. | 149 |
| 6.4 | Input parameters for the model. | 150 |
| 6.5 | Parameters for the permeability relationship, Eq.(6.8). [a]Ref. Shi and Durucan (2004a, 2006); Shi et al. (2008). | 151 |
| 7.1 | Parameter used for the model evaluation. | 180 |
| 8.1 | Salts used for the experiments, together with the corresponding obtained relative humidities, water vapor pressure and the moisture level of the coal sample. | 197 |

# Bibliography

Arri, L. E., Yee, D., Morgan, W. D., Jeansonne, M. W., 1992. Modeling coalbed methane production with binary gas sorption. SPE Paper 24363. Presented at the SPE Rocky Mountain Regional Meeting, Casper, Wyoming, May 18-21.

Atkinson, J., 2007. The Mechanics of Soils and Foundations, 2nd Edition. Taylor and Francis, London.

Bachu, S., 2008. $CO_2$ storage in geological media: Role, means, status and barriers to deployment. Prog. Energy Combust. Sci. 34 (2), 254–273.

Bachu, S., Gunter, W. D., Perkins, E. H., 1994. Aquifer disposal of $CO_2$: Hydrodynamic and mineral trapping. Energy Convers. Manage. 35 (4), 269–279.

Bae, J. S., Bhatia, S. K., 2006. High-pressure adsorption of methane and carbon dioxide on coal. Energy Fuels 20 (6), 2599–2607.

Berg, R. R., 1975. Capillary pressure in stratigraphic traps. Am. Assoc. Pet. Geol. Bull. 59 (6), 939–956.

Bonavoglia, B., Storti, G., Morbidelli, M., Rajendran, A., Mazzotti, M., 2006. Sorption and swelling of semicrystalline polymers in supercritical $CO_2$. J. Polym. Sci. Part B: Polym. Phys. 44 (11), 1531–1546.

Brace, W. F., Walsh, J. B., Frangos, W. T., 1968. Permeability of granite under high pressure. J. Geophys. Res. 73 (6), 2225–2236.

Bromhal, G. S., Neal Sams, W., Jikich, S., Ertekin, T., Smith, D. H., 2005. Simulation of $CO_2$ sequestration in coal beds: The effects of sorption isotherms. Chem. Geol. 217 (3-4), 201–211.

Busch, A., Gensterblum, Y., Krooss, B. M., 2003. Methane and $CO_2$ sorption and desorption measurements on dry Argonne Premium coals: pure components and mixtures. Int. J. Coal Geol. 55 (2-4), 205–224.

Busch, A., Gensterblum, Y., Krooss, B. M., 2007. High-pressure sorption of nitrogen, carbon dioxide, and their mixtures on Argonne Premium coals. Energy Fuels 21 (3), 1640–1645.

Busch, A., Gensterblum, Y., Krooss, B. M., Littke, R., 2004. Methane and carbon dioxide adsorption-diffusion experiments on coal: upscaling and modeling. Int. J. Coal Geol. 60 (2-4), 151–168.

Busch, A., Gensterblum, Y., Krooss, B. M., Siemons, N., 2006. Investigation of high-pressure selective adsorption/desorption $CO_2$ and $CH_4$ on coals: An experimental study. Int. J. Coal Geol. 66 (1-2), 53–68.

Bustin, A. M. M., Bustin, R. M., 2008. Coal reservoir saturation: Impact of temperature and pressure. AAPG Bull. 92 (1), 77–86.

Bustin, R. M., Cui, X. J., Chikatamarla, L., 2008. Impacts of volumetric strain on $CO_2$ sequestration in coals and enhanced $CH_4$ recovery. AAPG Bull. 92 (1), 15–29.

# Bibliography

Ceglarska-Stefanska, G., Czaplinski, A., 1993. Correlation between sorption and dilatometric processes in hard coals. Fuel 72 (3), 413–417.

Ceglarska-Stefanska, G., Zarebska, K., 2005. Sorption of carbon dioxide-methane mixtures. Int. J. Coal Geol. 62 (4), 211–222.

Chaback, J. J., Morgan, W. D., Yee, D., 1996. Sorption of nitrogen, methane, carbon dioxide and their mixtures on bituminous coals at in-situ conditions. Fluid Phase Equilib. 117 (1-2), 289–296.

Chadwick, R. A., Zweigel, P., Gregersen, U., Kirby, G. A., Holloway, S., Johannessen, P. N., 2004. Geological reservoir characterization of a $CO_2$ storage site: The Utsira Sand, Sleipner, northern North Sea. Energy 29 (9-10), 1371–1381.

Clarkson, C. R., Bustin, R. M., 2000. Binary gas adsorption/desorption isotherms: effect of moisture and coal composition upon carbon dioxide selectivity over methane. Int. J. Coal Geol. 42 (4), 241–271.

Close, J. C., 1993. Natural fractures in coal. In: Law, B., Rice, D. (Eds.), Hydrocarbons from Coal (AAPG Studies in Geology) ♯ 38. pp. 119–132.

Cui, X. J., Bustin, R. M., Chikatamarla, L., 2007. Adsorption-induced coal swelling and stress: Implications for methane production and acid gas sequestration into coal seams. J. Geophys. Res. [Solid Earth] 112 (B10202), 1–16.

Day, S., Duffy, G., Sakurovs, R., Weir, S., 2008a. Effect of coal properties on $CO_2$ sorption capacity under supercritical conditions. Int. J. Greenhouse Gas Control 2 (3), 342–352.

Day, S., Fry, R., Sakurovs, R., 2008b. Swelling of australian coals in supercritical $CO_2$. Int. J. Coal Geol. 74 (1), 41–52.

Day, S., Sakurovs, R., Weir, S., 2008c. Supercritical gas sorption on moist coals. Int. J. Coal Geol. 74 (3-4), 203–214.

DeGance, A. E., Morgan, W. D., Yee, D., 1993. High-pressure adsorption of methane, nitrogen and carbon-dioxide on coal substrates. Fluid Phase Equilib. 82, 215–224.

Diamond, W. P., Schatzel, S. J., 1998. Measuring the gas content of coal: A review. Int. J. Coal Geol. 35 (1-4), 311–331.

Diebold, P., 1988. Der Nordschweizer Permokarbon-Trog und die Steinkohlenfrage der Nordschweiz. Vierteljehrschrift der Naturforschenden Gesellschaft in Zürich 133 (1), 143–174.

Dubinin, M. M., Stoeckli, H. F., 1980. Homogeneous and heterogeneous micropore structures in carbonaceous adsorbents. J. Colloid Interface Sci. 75 (1), 34–42.

Durucan, S., Ahsan, M., Shi, J. Q., 2008. Matrix shrinkage and swelling characteristics of European coals. In: Proceedings of the 9th International Conference on Greenhouse Gas Control Technologies. Washington DC, USA, November 16-20.

Durucan, S., Shi, J.-Q., 2009. Improving the $CO_2$ well injectivity and enhanced coalbed methane production performance in coal seams. Int. J. Coal Geol. 77 (1-2), 214–221.

Fisher, G. J., 1992. The determination of permeability and storage capacity: pore pressure and oscillation method. In: Evans, B., Tengfong, W. (Eds.), Fault Mechanics and Transport Properties of Rocks. Vol. 51. Academic Press, London, pp. 187–211.

Fitzgerald, J. E., Pan, Z., Sudibandriyo, M., Robinson Jr., R. L., Gasem, K. A. M., Reeves, S., 2005. Adsorption of methane, nitrogen, carbon dioxide and their mixtures on wet Tiffany coal. Fuel 84 (18), 2351–2363.

Fitzgerald, J. E., Robinson Jr., R. L., Gasem, K. A. M., 2006. Modeling high-pressure adsorption of gas mixtures on activated carbon and coal using a simplified local-density model. Langmuir 22 (23), 9610–9618.

Fornstedt, T., Zhong, G., Bensetiti, Z., Guiochon, G., 1996. Experimental and theoretical study of the adsorption behavior and mass transfer kinetics of propranolol enantiomers on cellulase protein as the selector. Anal. Chem. 68 (14), 2370–2378.

Gao, W. H., Butler, D., Tomasko, D. L., 2004. High-pressure adsorption of $CO_2$ on NaY zeolite and model prediction of adsorption isotherms. Langmuir 20 (19), 8083–8089.

Gentzis, T., 2000. Subsurface sequestration of carbon dioxide - an overview from an Alberta (Canada) perspective. Int. J. Coal Geol. 43 (1-4), 287–305.

Gentzis, T., Deisman, N., Chalaturnyk, R. J., 2007. Geomechanical properties and permeability of coals from the Foothills and Mountain regions of western Canada. Int. J. Coal Geol. 69 (3), 153–164.

Gilman, A., Beckie, R., 2000. Flow of coal-bed methane to a gallery. Transp. Porous Media 41 (1), 1–16.

Goodman, A. L., Busch, A., Bustin, R. M., Chikatamarla, L., Day, S., Duffy, G. J., Fitzgerald, J. E., Gasem, K. A. M., Gensterblum, Y., Hartman, C., Jing, C., Krooss, B. M., Mohammed, S., Pratt, T., Robinson, R. L., Romanov, V., Sakurovs, R., Schroeder, K., White, C. M., 2007. Inter-laboratory comparison II: $CO_2$ isotherms measured on moisture-equilibrated Argonne Premium coals at 55°C and up to 15 MPa. Int. J. Coal Geol. 72 (3-4), 153–164.

Goodman, A. L., Busch, A., Duffy, G. J., Fitzgerald, J. E., Gasem, K. A. M., Gensterblum, Y., Krooss, B. M., Levy, J., Ozdemir, E., Pan, Z., Robinson Jr., R. L., Schroeder, K., Sudibandriyo, M., White, C. M., 2004. An inter-laboratory comparison of $CO_2$ isotherms measured on Argonne Premium coal samples. Energy Fuels 18 (4), 1175–1182.

Gray, I., 1987. Reservoir engineering in coal seams: Part 1 - the physical process of gas storage and movement in coal seams. Spe Reservoir Engineering SPE Paper 12514, 28–34.

Gruszkiewicz, M. S., Naney, M. T., Blencoe, J. G., Cole, D. R., Pashin, J. C., Carroll, R. E., 2009. Adsorption kinetics of $CO_2$, $CH_4$, and their equimolar mixture on coal from the Black Warrior Basin, West-Central Alabama. Int. J. Coal Geol. 77 (1-2), 23–33.

Gunter, W. D., Mavor, M. J., Robinson, J. R., 2004. $CO_2$ storage and enhanced methane production: field testing at the Fenn-Big Valley, Alberta, Canada, with application. In: Proceedings of the 7th International Conference on Greenhouse Gas Control Technologies. Vancouver, Canada, September 5-9.

Gunter, W. D., Perkins, E. H., McCann, T. J., 1993. Aquifer disposal of $CO_2$-rich gases: Reaction design for added capacity. Energy Convers. Manage. 34 (9-11), 941–948.

Harpalani, S., Chen, G., 1995. Estimation of changes in fracture porosity of coal with gas emission. Fuel 74 (10), 1491–1498.

Harpalani, S., Chen, G., 1997. Influence of gas production induced volumetric strain on permeability of coal. Geotech. Geol. Eng. 15 (4), 303–325.

# Bibliography

Harpalani, S., Schraufnagel, R. A., 1990. Shrinkage of coal matrix with release of gas and its impact on permeability of coal. Fuel 69 (5), 551–556.

Hesse, M. A., Orr, F. M., Tchelepi, H. A., 2008. Gravity currents with residual trapping. J. Fluid Mech. 611, 35–60.

Hildenbrand, A., Krooss, B. M., Busch, A., Gaschnitz, R., 2006. Evolution of methane sorption capacity of coal seams as a function of burial history – a case study from the Campine Basin, NE Belgium. Int. J. Coal Geol. 66 (3), 179–203.

Hildenbrand, A., Schlmer, S., Krooss, B. M., 2002. Gas breakthrough experiments on fine-grained sedimentary rocks. Geofluids 2 (1), 3–23.

Hildenbrand, A., Schlomer, S., Krooss, B. M., Littke, R., 2004. Gas breakthrough experiments on pelitic rocks: comparative study with $N_2$, $CO_2$ and $CH_4$. Geofluids 4 (1), 61–80.

Hocker, T., Rajendran, A., Mazzotti, M., 2003. Measuring and modeling supercritical adsorption in porous solids. Carbon dioxide on 13X zeolite and on silica gel. Langmuir 19 (4), 1254–1267.

Huang, H., Wang, S., Wang, K., Klein, M. T., Calkins, W. H., Davis, A., 1999. Thermogravimetric and Rock-Eval studies of coal properties and coal rank. Energy Fuels 13 (2), 396–400.

Humayun, R., Tomasko, D. L., 2000. High-resolution adsorption isotherms of supercritical carbon dioxide on activated carbon. AIChE J. 46 (10), 2065–2075.

Hutson, N. D., Yang, R. T., 1997. Theoretical basis for the Dubinin-Radushkevitch (D-R) adsorption isotherm equation. Adsorption 3 (3), 189–195.

IEA, 2008. Key world energy statistics. www.iea.org.

IPCC, 2005. IPCC Special Report on Carbon Dioxide Capture and Storage. Prepared by Working Group III of the Intergovernmental Panel on Climate Change. [Metz, B., O. Davidson, H. C. de Coninck, M. Loos, and L. A. Meyer (eds.)]. Cambridge University Press, Cambridge, United Kingdom and New York, NY, USA.

IPCC, 2007. Climate Change 2007: Synthesis Report. Contribution of Working Groups I, II and III to the Fourth Assessment Report of the Intergovernmental Panel on Climate Change. [Core Writing Team, Pachauri, R.K and Reisinger, A. (eds)]. IPCC, Geneva, Switzerland.

Jakubov, T. S., Mainwaring, D. E., 2002. Adsorption-induced dimensional changes of solids. PCCP 4 (22), 5678–5682.

Jessen, K., Tang, G.-Q., Kovscek, A., 2008. Laboratory and simulation investigation of enhanced coalbed methane recovery by gas injection. Transp. Porous Media 73 (2), 141–159.

Juanes, R., Spiteri, E., Orr Jr., F., Blunt, M., 2006. Impact of relative permeability hysteresis on geological $CO_2$ storage. Water Resour. Res. 42, 1–13.

Karacan, C. O., 2003. Heterogeneous sorption and swelling in a confined and stressed coal during $CO_2$ injection. Energy Fuels 17 (6), 1595–1608.

Keller, J. U., Staudt, R., 2005. Gas adsorption equilibria: experimental methods and adsorption isotherms. Springer Science+Business Media, Inc., New York.

Korre, A., Shi, J. Q., Imrie, C., Grattoni, C., Durucan, S., 2007. Coalbed methane reservoir data and simulator parameter uncertainty modelling for $CO_2$ storage performance assessment. Int. J. Greenhouse Gas Control 1 (4), 492–501.

Krooss, B. M., van Bergen, F., Gensterblum, Y., Siemons, N., Pagnier, H. J. M., David, P., 2002. High-pressure methane and carbon dioxide adsorption on dry and moisture-equilibrated Pennsylvanian coals. Int. J. Coal Geol. 51 (2), 69–92.

Kurniawan, Y., Bhatia, S. K., Rudolph, V., 2006. Simulation of binary mixture adsorption of methane and $CO_2$ at supercritical conditions in carbons. AIChE J. 52 (3), 957–967.

Larsen, J. W., 2004. The effects of dissolved $CO_2$ on coal structure and properties. Int. J. Coal Geol. 57 (1), 63–70.

Larsen, J. W., Flowers, R. A., Hall, P. J., Carlson, G., 1997. Structural rearrangement of strained coals. Energy Fuels 11 (5), 998–1002.

Lenormand, R., Zarcone, C., Sarr, A., 1983. Mechanism of the displacement of one fluid by another in a network of capillary ducts. J. Fluid Mech. 135, 337–353.

Levine, J., 1996. Model study of the influence of matrix shrinkage on absolute permeability of coal bed reservoirs. In: Gayer, R., Harris, I. (Eds.), Coalbed Methane and Coal Geology. Vol. 109. Geological Society Special Publication, London, pp. 197–212.

Levine, J. R., 1993. Coalification: The evolution of coal as source rock and reservoir rock for oil and gas. In: Law, B., Rice, D. (Eds.), Hydrocarbons from Coal (AAPG Studies in Geology) ♯ 38. pp. 39–77.

Li, S., Dong, M., Li, Z., Huang, S., Qing, H., Nickel, E., 2005. Gas breakthrough pressure for hydrocarbon reservoir seal rocks: implications for the security of long-term $CO_2$ storage in the Weyburn field. Geofluids 5 (4), 326–334.

Li, Z. W., Dong, M. Z., Li, S. L., Huang, S., 2006. $CO_2$ sequestration in depleted oil and gas reservoirs - caprock characterization and storage capacity. Energy Convers. Manage. 47 (11-12), 1372–1382.

Litynski, J., Plasynski, S., Spangler, L., Finley, R., Steadman, E., Ball, D., Nemeth, J. K., McPherson, B., Myer, L., 2008. U. S. Department of Energy's regional carbon sequestration partnership program: overview. In: Proceedings of the 9th International Conference on Greenhouse Gas Control Technologies. Washington DC, USA, November 16-20.

Mahajan, O. P., 1991. $CO_2$ surface area of coals: The 25-year paradox. Carbon 29 (6), 735–742.

Malbrunot, P., Vidal, D., Vermesse, J., Chahine, R., Bose, T. K., 1997. Adsorbent helium density measurement and its effect on adsorption isotherms at high pressure. Langmuir 13 (3), 539–544.

Mares, T. E., Moore, T. A., Moore, C. R., 2009. Uncertainty of gas saturation estimates in a subbituminous coal seam. Int. J. Coal Geol. 77 (3-4), 320–327.

Marschall, P., Horseman, S., Gimmi, T., 2005. Characterisation of gas transport properties of Opalinus clay, a potential host rock formation for radioactive waste disposal. Oil & Gas Science and Technology 60 (1), 121–139.

Massarotto, P., Golding, S. D., Iyer, R., Bae, J. S., Rudolph, V., 2007. Adsorption, Porosity and Permeability Effects of $CO_2$ Geosequestration in Permian Coals. Presented at the International Coalbed Methane Symposium, Tuscaloosa, Alabama U.S.A, May 23-24.

# Bibliography

Mastalerz, M., Gluskoter, H., Rupp, J., 2004. Carbon dioxide and methane sorption in high volatile bituminous coals from Indiana, USA. Int. J. Coal Geol. 60 (1), 43–55.

Mazumder, S., Karnik, A., Wolf, K. H., 2006a. Swelling of coal in response to $CO_2$ sequestration for ECBM and its effect on fracture permeability. SPE Journal 11 (3), 390–398.

Mazumder, S., van Hemert, P., Busch, A., Wolf, K.-H. A. A., Tejera-Cuesta, P., 2006b. Flue gas and pure $CO_2$ sorption properties of coal: A comparative study. Int. J. Coal Geol. 67 (4), 267–279.

Mazumder, S., Wolf, K. H., 2008. Differential swelling and permeability change of coal in response to $CO_2$ injection for ECBM. Int. J. Coal Geol. 74 (2), 123–138.

Mazzotti, M., Pini, R., Storti, G., 2009. Enhanced coal bed methane recovery. J. Supercrit. Fluids 47 (3), 619–617.

McCartney, J. T., Teichmller, M., 1972. Classification of coals according to degree of coalification by reflectance of the vitrinite component. Fuel 51 (1), 64–68.

McElhiney, J. E., Paul, G. W., Young, G. B. C., McCartney, J. A., 1993. Reservoir engineering aspects of coalbed methane. In: Law, B., Rice, D. (Eds.), Hydrocarbons from Coal, AAPG Studies in Geology 38. American Association of Petroleum Geologists, Tulsa, Oklahoma, pp. 361–372.

Milewska-Duda, J., 1987. Polymeric model of coal in the light of sorptive investigations. Fuel 66 (11), 1570–1573.

MIT, 2007. The future of coal. Options for a carbon-constrained world. An interdisciplinary MIT study. Massachusetts Institute of Technology.

Montel, F., Caillet, G., Pucheu, A., Caltagirone, J. P., 1993. Diffusion model for predicting reservoir gas losses. Mar. Pet. Geol. 10 (1), 51–57.

Morbidelli, M., Servida, A., Storti, G., 1983. Application of the orthogonal collocation method to some chemical engineering problems. Ing. Chim. Ital. 19 (5-6), 46–60.

Mukhopadhyay, P. K., Hatcher, P. G., 1993. Composition of coal. In: Law, B., Rice, D. (Eds.), Hydrocarbons from Coal (AAPG Studies in Geology) ♯ 38. pp. 79–118.

Murata, K., El-Merraoui, M., Kaneko, K., 2001. A new determination method of absolute adsorption isotherm of supercritical gases under high pressure with a special relevance to density-functional theory study. J. Chem. Phys. 114 (9), 4196–4205.

Nagra, 1989. Sondierbohrung Weiach Untersuchungsbericht. Technischer Bericht NTB 88-08.

Nagra, 2002. Projekt Opalinuston - Synthese der geowissenschaftlichen Untersuchungsergebnisse. Technischer Bericht NTB 02-03.

Neuzil, C. E., 2003. Hydromechanical coupling in geologic processes. Hydrol. J. 11 (1), 41–83.

NIST, 2008. NIST Chemistry WebBook. http://webbook.nist.gov/chemistry.

Orr, F. M., 2007. Theory of gas injection processes. Tie-Line Publications, Copenhagen.

# Bibliography

Ottaway, M., 1982. Use of thermogravimetry for proximate analysis of coals and cokes. Fuel 61 (8), 713–716.

Ottiger, S., Pini, R., Storti, G., Mazzotti, M., 2008a. Competitive adsorption equilibria of $CO_2$ and $CH_4$ on a dry coal. Adsorption 14 (4-5), 539–556.

Ottiger, S., Pini, R., Storti, G., Mazzotti, M., 2008b. Measuring and modeling the competitive adsorption of $CO_2$, $CH_4$ and $N_2$ on a dry coal. Langmuir 24 (17), 9531–9540.

Ottiger, S., Pini, R., Storti, G., Mazzotti, M., Bencini, R., Quattrocchi, F., Sardu, G., Deriu, G., 2006. Adsorption of pure carbon dioxide and methane on dry coal from the Sulcis Coal Province (SW Sardinia, Italy). Environ. Prog. 25 (4), 355–364.

Ozdemir, E., Morsi, B. I., Schroeder, K., 2003. Importance of volume effects to adsorption isotherms of carbon dioxide on coals. Langmuir 19 (23), 9764–9773.

Ozdemir, E., Morsi, B. I., Schroeder, K., 2004. $CO_2$ adsorption capacity of Argonne Premium coals. Fuel 83 (7-8), 1085–1094.

Palmer, I., Mansoori, J., 1998. How permeability depends on stress and pore pressure in coalbeds: A new model. Spe Reserv. Eval. Eng. 1 (6), 539–544.

Pan, Z., Connell, L. D., 2007. A theoretical model for gas adsorption-induced coal swelling. Int. J. Coal Geol. 69 (4), 243–252.

Pekot, L. J., Reeves, S. R., 2003. Modeling the effects of matrix shrinkage and differential swelling on coalbed methane recovery and carbon sequestration. In: International Coalbed Methane Symposium, paper 0328. University of Alabama, Tuscaloosa, Alabama, May 59.

Peng, D.-Y., Robinson, D. B., 1976. A new two-constant equation of state. Ind. Eng. Chem. Fund. 15 (1), 59–64.

Pini, R., Ottiger, S., Burlini, L., Storti, G., Mazzotti, M., 2009. Role of adsorption and swelling on the dynamics of gas injection in coal. J. Geophys. Res. [Solid Earth] 114, B04203.

Pini, R., Ottiger, S., Rajendran, A., Storti, G., Mazzotti, M., 2006. Reliable measurement of near-critical adsorption by gravimetric method. Adsorption 12 (5-6), 393–403.

Pini, R., Storti, G., Mazzotti, M., Tai, H., Shakesheff, K. M., Howdle, S. M., 2007. Sorption and swelling of poly(DL-lactic acid) and poly(lactic-co-glycolic acid) in supercritical $CO_2$. Macromolecular Symposia 259 (1), 197–202.

Pini, R., Storti, G., Mazzotti, M., Tai, H., Shakesheff, K. M., Howdle, S. M., 2008. Sorption and swelling of poly(DL-lactic acid) and poly(lactic-co-glicolyc acid) in supercritical $CO_2$: an experimental and modeling study. J. Polym. Sci. Part B: Polym. Phys. 46 (5), 483–496.

Rajendran, A., Bonavoglia, B., Forrer, N., Storti, G., Mazzotti, M., Morbidelli, M., 2005. Simultaneous measurement of swelling and sorption in a supercritical $CO_2$-poly(methyl methacrylate) system. Ind. Eng. Chem. Res. 44 (8), 2549–2560.

Reeves, S. R., 2004. The Coal-Seq project: Key results from field, laboratory, and modeling studies. In: Proceedings of the 7th International Conference on Greenhouse Gas Control Technologies. Vancouver, Canada, September 5-9.

Reid, R. C., Prausnitz, J. M., Poling, B. E., 1987. The Properties of Gases and Liquids, 4th Edition. McGraw-Hill, New York.

# Bibliography

Reiss, L. H., 1980. Reservoir engineering en milieu fissuré. Editions Technip, Paris.

Reucroft, P. J., Sethuraman, A. R., 1987. Effect of pressure on carbon-dioxide induced coal swelling. Energy Fuels 1 (1), 72–75.

Rhee, H.-K., Aris, R., Amundson, N. R., 2001. First-order partial differential equations - Volume 1. Dover Publications, Inc., New YOrk.

Romanov, V., Soong, Y., Schroeder, K., 2006. Volumetric effects in coal sorption capacity measurements. Chem. Eng. Technol. 29 (3), 368–374.

Ruthven, D. M., 1984. Principles of Adsorption and Adsorption Processes. Wiley, New York.

Saghafi, A., Faiz, M., Roberts, D., 2007. $CO_2$ storage and gas diffusivity properties of coals from Sydney Basin, Australia. Int. J. Coal Geol. 70 (1-3), 240–254.

Sakurovs, R., Day, S., Weir, S., 2008a. Causes and consequences of errors in determining sorption capacity of coals for carbon dioxide at high pressure. Int. J. Coal Geol. 77 (1-2), 16–22.

Sakurovs, R., Day, S., Weir, S., Duffy, G., 2007. Application of a modified Dubinin- Radushkevich equation to adsorption of gases by coals under supercritical conditions. Energy Fuels 21 (2), 992–997.

Sakurovs, R., Day, S., Weir, S., Duffy, G., 2008b. Temperature dependence of sorption of gases by coals and charcoals. Int. J. Coal Geol. 73 (3-4), 250–258.

Sams, W. N., Bromhal, G., Jikich, S., Ertekin, T., Smith, D. H., 2005. Field-project designs for carbon dioxide sequestration and enhanced coalbed methane production. Energy Fuels 19 (6), 2287–2297.

Scherer, G. W., 1986. Dilatation of porous-glass. J. Am. Ceram. Soc. 69 (6), 473–480.

Schlomer, S., Krooss, B. M., 1997. Experimental characterisation of the hydrocarbon sealing efficiency of cap rocks. Mar. Pet. Geol. 14 (5), 563–578.

Seidle, J. P., Jeansonne, M. W., Erickson, D. J., 1992. Application of matchstick geometry to stress dependent permeability in coals. In: SPE Paper 24361. Presented at the SPE Rocky Mountain Regional Meeting, Casper, Wyoming U.S.A, May 18-21.

Seto, C., 2007. Analytical theory for two-phase, multicomponent flow in porous media with adsorption. Stanford University, PhD Dissertation.

Seto, C. J., Jessen, K., Orr Jr., F. M., 2006. A four-component, two-phase flow model for $CO_2$ storage and enhanced coalbed methane recovery. In: SPE Paper 102376. Presented at the SPE Annual Technical Conference and Exhibition in San Antonio, Texas U.S.A, September 24-27.

Shi, J. Q., Durucan, S., 2003. A bidisperse pore diffusion model for methane displacement desorption in coal by $CO_2$ injection. Fuel 82 (10), 1219–1229.

Shi, J. Q., Durucan, S., 2004a. Drawdown induced changes in permeability of coalbeds: A new interpretation of the reservoir response to primary recovery. Transp. Porous Media 56 (1), 1–16.

Shi, J. Q., Durucan, S., 2004b. A numerical simulation study of the allison unit $CO_2$-ECBM pilot: the impact of matrix shrinkage and swelling on ECBM production and $CO_2$ injectivity,. In: Proceedings of the 7th International Conference on Greenhouse Gas Control Technologies. Vancouver, Canada, September 5-9.

Shi, J. Q., Durucan, S., 2005a. Gas storage and flow in coalbed reservoirs: Implementation of a bidisperse pore model for gas diffusion in a coal matrix. Spe Reserv. Eval. Eng. 8 (2), 169–175.

Shi, J. Q., Durucan, S., 2005b. A model for changes in coalbed permeability during primary and enhanced methane recovery. Spe Reserv. Eval. Eng. 8 (4), 291–299.

Shi, J. Q., Durucan, S., 2006. The assessment of horizontal well option for $CO_2$ storage and ECBM recovery in unmineable thin seams: Pure $CO_2$ vs $CO_2$ enriched flue gas. In: Proceedings of the 8th International Conference on Greenhouse Gas Control Technologies. Trondheim, Norway, June 19-22.

Shi, J.-Q., Durucan, S., Fujioka, M., 2008. A reservoir simulation study of $CO_2$ injection and $N_2$ flooding at the Ishikari coalfield $CO_2$ storage pilot project, Japan. Int. J. Greenhouse Gas Control 2 (1), 47–57.

Shimada, S., Li, H. Y., Oshima, Y., Adachi, K., 2005. Displacement behavior of $CH_4$ adsorbed on coals by injecting pure $CO_2$, $N_2$, and $CO_2$-$N_2$ mixture. Environ. Geol. 49 (1), 44–52.

Siemons, N., Busch, A., 2007. Measurement and interpretation of supercritical $CO_2$ sorption on various coals. Int. J. Coal Geol. 69 (4), 229–242.

Sircar, S., 1999. Gibbsian surface excess for gas adsorption - revisited. Ind. Eng. Chem. Res. 38 (10), 3670–3682.

Sircar, S., 2001. Measurement of gibbsian surface excess. AIChE J. 47 (5), 1169–1176.

Smith, D. H., Bromhal, G., Sams, W. N., Jikich, S., Ertekin, T., 2005. Simulating carbon dioxide sequestration/ECBM production in coal seams: Effects of permeability anisotropies and the diffusion-time constant. Spe Reserv. Eval. Eng. 8 (2), 156–163.

Span, R., Wagner, W., 1996. A new equation of state for carbon dioxide covering the fluid region from the triple-point temperature to 1100 K at pressures up to 800 MPa. J. Phys. Chem. Ref. Data 25 (6), 1509–1596.

Spycher, N., Pruess, K., Ennis-King, J., 2003. $CO_2$-$H_2O$ mixtures in the geological sequestration of $CO_2$. i. assessment and calculation of mutual solubilities from 12 to 100°C and up to 600 bar. Geochim. Cosmochim. Acta 67 (16), 3015–3031.

St. George, J. D., Barakat, M. A., 2001. The change in effective stress associated with shrinkage from gas desorption in coal. Int. J. Coal Geol. 45 (2-3), 105–113.

Stevenson, M. D., Pinczewski, W. V., Somers, M. L., Bagio, S. E., 1991. Adsorption/desorption of multicomponent gas mixtures at in-seam conditions. In: SPE Paper 23026. Presented at the SPE Asia-Pacific Conference, Perth, Western Australia, November 4-7.

Sudibandriyo, M., Fitzgerald, J. E., Pan, Z., Robinson Jr., R. L., Gasem, K. A. M., 2003a. Extension of the Ono-Kondo lattice model to high-pressure mixture adsorption. In: Proceedings of the AIChE Spring National Meeting. New Orleans, LA, March 30-April 3.

Sudibandriyo, M., Pan, Z. J., Fitzgerald, J. E., Robinson, R. L., Gasem, E. A. M., 2003b. Adsorption of methane, nitrogen, carbon dioxide, and their binary mixtures on dry activated carbon at 318.2 K and pressures up to 13.6 MPa. Langmuir 19 (13), 5323–5331.

Tang, G. Q., Jessen, K., Kovscek, A. R., 2005. Laboratory and Simulation Investigation of Enhanced Coalbed Methane Recovery by Gas

# Bibliography

Injection. SPE Paper 95947. Presented at the SPE Annual Technical Conference and Exhibition, Dallas, Texas U.S.A, October 8-12.

Toribio, M., Oshima, Y., Shimada, S., Pini, R., Ottiger, S., Storti, G., Mazzotti, M., 2005. Adsorption measurement of supercritical $CO_2$ on coal. In: Proceedings of the International Conference on Coal Science and Technology (ICCS&T). Okinawa, Japan, October 9-14.

Totsis, T. T., Patel, H., Najafi, B. F., Racherla, D., Knackstedt, M. A., Sahimi, M., 2004. Overview of laboratory and modeling studies of carbon dioxide sequestration in coal beds. Ind. Eng. Chem. Res. 43, 2887–2901.

Ustinov, E. A., Do, D. D., Herbst, A., Staudt, R., Harting, P., 2002. Modeling of gas adsorption equilibrium over a wide range of pressure: A thermodynamic approach based on equation of state. J. Colloid Interface Sci. 250 (1), 49–62.

Van Bergen, F., Pagnier, H., Krzystolik, P., 2006. Field experiment of $CO_2$-ECBM in the Upper Silesian Basin of Poland. In: Proceedings of the 8th International Conference on Greenhouse Gas Control Technologies. Trondheim, Norway, June 19-22.

Van Krevelen, D. W., 1981. Coal : typology - chemistry - physics - constitution. In: Anderson, L. (Ed.), Coal Science and Technology. Elsevier Science Publishers, Amsterdam.

Viete, D. R., Ranjith, P. G., 2006. The effect of $CO_2$ on the geomechanical and permeability behaviour of brown coal: Implications for coal seam $CO_2$ sequestration. Int. J. Coal Geol. 66 (3), 204–216.

Viete, D. R., Ranjith, P. G., 2007. The mechanical behaviour of coal with respect to $CO_2$ sequestration in deep coal seams. Fuel 86 (17-18), 2667–2671.

Villadsen, J., Michelsen, M. L., 1978. Solution of Differential Equation Models by Polynomial Approximation. Prentice-Hall international series in the physical and chemical engineering science. Prentice-Hall, Englewood Cliffs, NJ.

Walker, Jr, P. L., Verma, S. K., Rivera-Utrilla, J., Khan, M. R., 1988. A direct measurement of expansion in coals and macerals induced by carbon dioxide and methanol. Fuel 67 (5), 719–726.

Wang, F. Y., Zhu, Z. H., Massarotto, P., Rudolph, V., 2007. Mass transfer in coal seams for $CO_2$ sequestration. AIChE J. 53 (4), 1028–1049.

Watts, N. L., 1987. Theoretical aspects of cap-rock and fault seals for single- and two-phase hydrocarbon columns. Mar. Pet. Geol. 4 (4), 274–307.

Wei, X. R., Wang, G. X., Massarotto, P., Golding, S. D., Rudolph, V., 2007a. Numerical simulation of multicomponent gas diffusion and flow in coals for $CO_2$ enhanced coalbed methane recovery. Chem. Eng. Sci. 62 (16), 4193–4203.

Wei, X. R., Wang, G. X., Massarotto, P., Rudolph, V., Golding, S. D., 2007b. Modeling gas displacement kinetics in coal with Maxwell-Stefan diffusion theory. AIChE J. 53 (12), 3241–3252.

White, C. M., Smith, D. H., Jones, K. L., Goodman, A. L., Jikich, S. A., LaCount, R. B., DuBose, S. B., Ozdemir, E., Morsi, B. I., Schroeder, K. T., 2005. Sequestration of carbon dioxide in coal with enhanced coalbed methane recovery - a review. Energy Fuels 19 (3), 659–724.

Wong, S., Law, D., Deng, X., Robinson, J., Kadatz, B., Gunter, W. D., Jianping, Y., Sanli, F., Zhiqiang, F., 2007. Enhanced coalbed methane and $CO_2$ storage in anthracitic coals-micro-pilot test at South Qinshui, Shanxi, China. Int. J. Greenhouse Gas Control 1 (2), 215–222.

Wong, S., Law, D., Deng, X., Robinson, J., Kadatz, B., Gunter, W. D., Ye, J., Feng, S., Fan, Z., 2006. Enhanced coalbed methane - micropilot test at South Qinshui, Shanxi, China. In: Proceedings of the 8th International Conference on Greenhouse Gas Control Technologies. Trondheim, Norway, June 19-22.

Yamaguchi, S., Ohga, K., Fujioka, M., Nako, M., Muto, S., 2006. Field experiment of Japan $CO_2$ geosequestration in coal seams project (JCOP). In: Proceedings of the 8th International Conference on Greenhouse Gas Control Technologies. Trondheim, Norway, June 19-22.

Yang, R. T., 1997. Gas separation by adsorption processes. Vol. 1 of Series on Chemical Engineering. Imperial College Press, London.

Yee, D., Seidele, J., Handson, 1993. Gas sorption on coal and measurements of gas content. In: Law, B., Rice, D. (Eds.), Hydrocarbons from Coal, AAPG Studies in Geology 38. American Association of Petroleum Geologists, Tulsa, Oklahoma, pp. 203–218.

Yu, H. G., Yuan, J., Guo, W. J., Cheng, J. L., Hu, Q. T., 2008a. A preliminary laboratory experiment on coalbed methane displacement with carbon dioxide injection. Int. J. Coal Geol. 73 (2), 156–166.

Yu, H. G., Zhou, L. L., Guo, W. L., Cheng, J., Hu, Q. T., 2008b. Predictions of the adsorption equilibrium of methane/carbon dioxide binary gas on coals using Langmuir and Ideal Adsorbed Solution theory under feed gas conditions. Int. J. Coal Geol. 73 (2), 115–129.

Zhu, J., Jessen, K., Kovscek, A. R., Orr Jr., F. M., 2003. Analytical theory of coalbed methane recovery by gas injection. Spe Journal 8 (4), 371–379.

Zhu, W. C., Liu, J., Sheng, J. C., Elsworth, D., 2007. Analysis of coupled gas flow and deformation process with desorption and Klinkenberg effects in coal seams. Int. J. Rock Mech. Min. Sci. 44 (7), 971–980.

Die VDM Verlagsservicegesellschaft sucht für wissenschaftliche Verlage abgeschlossene und herausragende

# Dissertationen, Habilitationen, Diplomarbeiten, Master Theses, Magisterarbeiten usw.

### für die kostenlose Publikation als Fachbuch.

Sie verfügen über eine Arbeit, die hohen inhaltlichen und formalen Ansprüchen genügt, und haben Interesse an einer honorarvergüteten Publikation?

Dann senden Sie bitte erste Informationen über sich und Ihre Arbeit per Email an *info@vdm-vsg.de*.

### Sie erhalten kurzfristig unser Feedback!

VDM Verlagsservicegesellschaft mbH
Dudweiler Landstr. 99
D - 66123 Saarbrücken

Telefon +49 681 3720 174
Fax     +49 681 3720 1749

### www.vdm-vsg.de

Die VDM Verlagsservicegesellschaft mbH vertritt

Printed by Books on Demand GmbH, Norderstedt / Germany